Edexcel
GCSE MATHEMATICS

PRACTICE BOOK: Higher

Gareth Cole   Peter Jolly   David Kent   Keith Pledger

Note: There are no Practice exercises for
Unit 11: Using and applying mathematics.

Edexcel
Success through qualifications

Heinemann

About this book

This book provides a substantial bank of additional exercises to complement those in the Edexcel GCSE Mathematics course textbook and offers a firm foundation for a programme of consolidation and homework.

Extra exercises are included for every topic covered in the course textbook, with the exception of the Using and applying and the Calculators and computers units.

Clear links to the course textbook exercises help you plan your use of the book:

Exercise 1.1	Links: $(1A - D)$ 1A – D

This exercise is linked to exercises 1A through 1D in the old edition of the course textbook.

This exercise is linked to exercises 1A through 1D in the new edition of the course textbook.

Please note that the answers to the questions are provided in a separate booklet, available free when you order a pack of 10 practice books. You can buy further copies direct from Heinemann Customer Services.

> **Which edition am I using?**
>
> The new editions of the *Edexcel GCSE Maths* core textbooks have yellow cover flashes saying "ideal for the 2001 specification". You can also use the old edition (no yellow cover flash) to help you prepare for your exam.

Also available from Heinemann:

Edexcel GCSE Mathematics: Higher

The Higher textbook provides a complete course for the Higher tier examination.

Revise for Edexcel GCSE: Higher

The Higher revision book provides a structured approach to pre-exam revision and helps target areas for review.

1 Exploring numbers

Exercise 1.1   **Links:** (*1A – D*) 1A – D

1 Find all the factors of:
   (a) 12   (b) 37   (c) 72
   (d) 616   (e) 49   (f) 2002

2 Work out the sum of the prime numbers between 20 and 30.

3 Which of these numbers are prime?
   (a) 3   (b) 23   (c) 27   (d) 1   (e) 91   (f) 97

4 Work out the value of:
   (a) $2^3 \times 3^2 \times 5$ (b) $3^3 \times 5^2$
   (c) $2^5 \times 3^2$ (d) $2^2 \times 5^2 \times 11^3$

5 Given that $2^n \times 3 = 24$, work out the value of n.

6 Write each of these numbers in prime factor form:
   (a) 36   (b) 72   (c) 39
   (d) 1440   (e) 200   (f) 2002

7 Find the Highest Common Factor of:
   (a) 6 and 15   (b) 48 and 72
   (c) 1500 and 504   (d) 99 and 3003

8 Find out whether these pairs of numbers are co-prime. Give a
   reason for each answer.
   (a) 15 and 42   (b) 16 and 45
   (c) 24 and 100   (d) 39 and 40

9 Write down any whole number which:
   (a) must be co-prime with 100   (b) is not co-prime with 17
   Give a reason for each answer.

10 Use Euclid's algorithm to find the HCF of:
   (a) 32 and 80   (b) 24 and 117   (c) 72 and 1960

11 Find the Lowest Common Multiple of:
   (a) 24 and 30   (b) 48 and 108
   (c) 60 and 45   (d) 36 and 49

12 Two numbers, n and m have a LCM of $n \times m$.
   What can you say about n and m?

13 Mr Khan has two flashing lamps.
He switches both of them on at
the same moment in time.
The first lamp flashes every 25 seconds.
The second lamp flashes every minute.
Work out the first three times after the lamps are switched on
when they will flash together.

Exercise 1.2   Links: (*1E*) 1E

1 Copy and complete this table for triangular numbers:

1st	2nd	3rd	4th	5th	6th	7th	8th	9th	10th
1	3	6	10						

2 Copy and complete this table of values for square numbers:

1st	2nd	3rd	4th	5th	6th	7th	8th	9th	10th
1	4	9	16						

3 Copy and complete this table of values for cube numbers:

1st	2nd	3rd	4th	5th	6th	7th	8th	9th	10th
1	8	27	64						

4 By considering the dot patterns, show that the sum of two
consecutive triangular numbers is a square number.

5 **(a)** Show that 25 and 144 are square numbers.
(b) Show that $25 + 144$ is also a square number.
(c) Find two other square numbers, both of which must be
greater than 25, which add together to give another square
number.

6 The 3rd square number is 9.
The 4th square number is 16.
The difference between these two consecutive square numbers is
$16 - 9 = 7$.
This result, 7, is an odd number.
By considering dot patterns for square numbers, show that the
difference between two consecutive square numbers is an odd
number.

7 The notation C_4 stands for the 4th cube number; i.e.
$C_4 = 4^3 = 4 \times 4 \times 4 = 64$. Show that $C_3 + C_4 + C_5 = C_6$.

8 $1 = 1$
$1 + 3 = 4$
$1 + 3 + 5 = 9$
$1 + 3 + 5 + 7 = 16$
(a) What do you notice is happening with this pattern of results?
(b) Use this to find the sum of the first 20 odd numbers.

9 $1 = 1$
$3 + 5 = 8$
$7 + 9 + 11 = 27$
$13 + 15 + 17 + 19 = 64$
(a) What do you notice is happening with this pattern of results?
(b) Write 1000 as the sum of consecutive odd numbers.

10 Find the next two numbers in the sequence:

$$1, \quad 9, \quad 36, \quad 100, \quad 225, \quad \ldots$$

11 Prove that the difference between the squares of two consecutive odd numbers is a multiple of 8.

Exercise 1.3   Links: (*1F, G, H*) 1F, G, H

Calculate:

1 **(a)** 3^3 **(b)** 5^2 **(c)** 10^3
    (d) $7^2 \times 2^3$ **(e)** 4^2 **(f)** 3^4
    (g) 7^5 **(h)** 12^3 **(i)** $3^2 \times 2^3$
    (j) $\left(\frac{1}{2}\right)^2$ **(k)** $(0.1)^2$ **(l)** 13^2
    (m) 15^4 **(n)** 2^{10} **(o)** $(4 \times 5)^3$

2 Write these numbers in index form:
    (a) 10 000   **(b)** 1 000 000
    (c) 225   **(d)** 343
    (e) 289   **(f)** 1024
    (g) 2197   **(h)** 625

3 **(a)** Show that:
       $1^2 = 1$ $11^2 = 121$ $111^2 = 12321$
    (b) Work out the values of:
       (i) 1111^2 **(ii)** 11111^2
    (c) By looking at the pattern in the above results, write down
       the value of:
       (i) 111111^2 **(ii)** 111111111^2

4 Which is the greater, and by how much?
   (a) 2^2 or 3^2 **(b)** 3^4 or 4^3

5 Work out each of the following.
   In each case leave your answer both in index form, and where possible, without using indices.
   (a) $7^2 \times 7^3$ **(b)** $10^3 \times 10^2$ **(c)** $2^5 \times 2^3$
   (d) $3^3 \times 3^2 \times 3$ **(e)** 12^0 **(f)** $5^3 \div 5^2$
   (g) $7^5 \div 7^3$ **(h)** $17^3 \times 17^0$ **(i)** $8^3 \times 4^2$
   (j) $7^2 \div 7^3$ **(k)** $15^3 \div 15$ **(l)** $15 \div 15^3$
   (m) $3 + 3^2 + 3^2$ **(n)** 2^{-1} **(o)** $(3^2)^3$
   (p) $(3^3)^2$ **(q)** $(7^2)^2$ **(r)** 10^{-2}
   (s) $(5^{-1})^2$ **(t)** $5^3 \div 5^{-2}$ **(u)** $(2^2)^{-3}$
   (v) $10^{-2} \div 10^{-3}$ **(w)** $(2^{-2})^{-3}$ **(x)** $5^3 \times 5^2 \div 5^{-2}$

6 Work out each of these, leaving your answer as an ordinary number.
   (a) $\dfrac{2^2 \times 2^3}{2}$ **(b)** $\dfrac{3^3 \times 3^2}{3^{-1}}$ **(c)** $\dfrac{(5^2)^3}{5^{-2}}$ **(d)** $\dfrac{(4^2)^3}{4^7}$

7 Without using a calculator, work out the exact value of:
   (a) $16^{\frac{1}{2}}$ **(b)** $10000^{\frac{1}{4}}$ **(c)** $32^{\frac{1}{5}}$
   (d) $64^{\frac{1}{3}}$ **(e)** $64^{-\frac{1}{3}}$ **(f)** $1000^{-\frac{1}{3}}$
   (g) $25^{\frac{3}{2}}$ **(h)** $49^{-\frac{3}{2}}$ **(i)** $169^{\frac{1}{2}}$
   (j) $169^{-\frac{1}{2}}$ **(k)** $169^{\frac{3}{2}}$ **(l)** $169^{-\frac{3}{2}}$
   (m) $(4^3)^{-\frac{2}{3}}$ **(n)** $1024^{-\frac{1}{10}}$ **(o)** $225^{-\frac{1}{2}}$
   (p) $8^{\frac{4}{3}}$ **(q)** $8^{-\frac{4}{3}}$ **(r)** $8^{\frac{1}{3}} \times 2^{-1}$
   (s) $27^{\frac{5}{3}}$ **(t)** $2^{\frac{1}{3}} \times 4^{\frac{1}{3}}$ **(u)** $25^{-\frac{3}{2}}$
   (v) $32^{-\frac{3}{5}}$ **(w)** $(4^2)^{-\frac{1}{4}}$ **(x)** $125^{\frac{4}{3}}$
   (y) $1\,000\,000^{-\frac{2}{3}}$

8 Given that $8^{\frac{1}{3}} \times p^n \times 21 = 1050$ where p is a prime number and n is a positive integer, find:
   (a) the value of p **(b)** the value of n

9 Express as a single fraction:
   (a) $(2^0 + 2^{-1} + 2^{-2} + 2^{-3})^2$
   (b) $(2^0 + 2^{-1} + 2^{-2} + 2^{-3})^{-1}$

Exercise 1.4   Links: (1I) 1I

1 Work out:
   (a) 2^6 **(b)** 2^{-6} **(c)** 2^{-3}
   (d) 10^3 **(e)** 10^{-3} **(f)** $2^3 \times 10^4$

2 Round 2^9 to the nearest 10.

3 Solve:
  (a) $2^{3n+1} = 32$ **(b)** $10^{5-2n} = 10\,000\,000$

Exercise 1.5   Links: (*1J*) 1J

1 The factors of 6 (excluding 6 itself) are 1, 2, 3.
The sum of these factors is 6.
Any number whose sum of factors equals the number itself is
called a **perfect number**.
(When you work out the sum of the factors you always exclude
the number itself.)
Find the next perfect number after 6.

2 The number 17 can be made by adding together multiples of 3 and 7

i.e. $17 = 1 \times 3 + 2 \times 7$

(By the word multiple we mean positive whole number
multiples.)
Find the largest number which **cannot** be made by adding
together multiples of 3 and 7.

3 The number 1998 can be written as $2 \times 3^n \times p$, where n is a
whole number and p is a prime number.
  (a) Work out the values of n and p.
  (b) Using your answers to part **(a)**, or otherwise, work out the
        factor of 1998 which is between 100 and 200.   [E]

4 The odd numbers, starting at 1, are set out in triangles.
The 4th and 5th triangles of odd numbers are shown below.

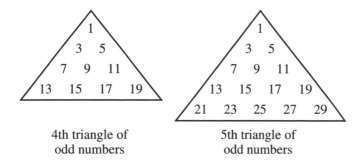

4th triangle of   5th triangle of
odd numbers   odd numbers

The sum of the numbers in the bottom row of the 4th triangle of
odd numbers is

  $13 + 15 + 17 + 19 = 64$

  (a) Write down, in terms of n, an expression for the sum of the
        numbers in the bottom row of the nth triangle of odd
        numbers.

(b) Use your expression to find the sum of the numbers in the bottom row of the 50th triangle of odd numbers.

(c) The sum of the numbers in the bottom row of the *n*th triangle of odd numbers is 3375.
Find the value of *n*.

(d) Investigate triangles of odd numbers to find an expression for the sum of all the numbers in the *n*th triangle of odd numbers.

(e) Prove that:

$$(1 + 2 + 3 + 4 + \ldots + n)^2 = 1^3 + 2^3 + 3^3 + 4^3 + \ldots + n^3$$

[E]

Exercise 1.6   Links: (*1K*) 1K

1 Work out:

(a) $1\frac{3}{5} + 2\frac{1}{4}$

(b) $3\frac{1}{7} + 4\frac{2}{3}$

(c) $5\frac{3}{4} - 1\frac{1}{3}$

(d) $2\frac{1}{4} \times 3\frac{2}{5}$

(e) $4\frac{1}{7} - 1\frac{3}{5}$

(f) $6\frac{3}{8} \div 2\frac{1}{2}$

(g) $5\frac{1}{2} \div 2\frac{1}{4}$

(h) $4\frac{1}{2} \times 2\frac{1}{3}$

(i) $3\frac{1}{2^x}\left(1\frac{1}{4} - \frac{3}{5}\right)$

(j) $\dfrac{4\frac{1}{3}}{2\frac{1}{2} - \frac{3}{4}}$

(k) $\dfrac{5\frac{1}{3} - 2\frac{1}{2}}{1\frac{1}{4}}$

(l) $\frac{4}{5} \div \frac{1}{4}$

2 Work out:

(a) $\dfrac{3\frac{1}{2} + 2\frac{1}{3}}{2\frac{1}{2} - \frac{3}{4}}$

(b) $\dfrac{4\frac{1}{3} - 1\frac{1}{2}}{5\frac{1}{2} - 2\frac{1}{4}}$

(c) $\dfrac{5\frac{2}{3} - 1\frac{4}{5}}{3\frac{1}{2} \times \frac{2}{3}}$

(d) $\dfrac{4\frac{1}{4}\left(7\frac{1}{2} - 3\frac{1}{5}\right)}{5\frac{2}{5} - 3\frac{1}{2}}$

2 Solving equations and inequalities

Solve the equations:

1 $2x + 1 = 9$ **2** $3x + 2 = 20$ **3** $4x = 24$

4 $8 + 5y = 18$ **5** $3b - 6 = 21$ **6** $7c - 8 = 20$

7 $9k + 1 = 26$ **8** $3k - 17 = -4$ **9** $5k + 11 - -20$

10 $2 - 3k = 17$ **11** $4 - 5m = 22$ **12** $3 = 7 - m$

13 $12 = 2p + 7$ **14** $17 = 18 - 5p$ **15** $2 - q = 19$

16 $3x + 1 = 2x + 4$ **17** $5x - 7 = 4x + 4$ **18** $16x + 1 = 12x + 1$

19 $2 - 3x = 2x + 7$ **20** $19x + 13 = 17x + 21$ **21** $15k - 3 = 4 + 6k$

22 $13 - 3p = 4 - 2p$ **23** $8 - 5q = 21 - 2q$ **24** $9 + 3q = 2q - 6$

25 $\frac{1}{3}x + 2 = 7$ **26** $\frac{1}{4}x + 5 = 16$ **27** $\frac{1}{5}x - 3 = 2$

28 $4 - \frac{1}{3}x = 6$ **29** $\frac{2}{3}x + 9 = 4$ **30** $5 - \frac{3}{4}x - 17$

1 The sum of three consecutive numbers is 45. Find the numbers.

2 In two years time I will be 3 times as old as I am now. How old am I?

3 If I multiply the number I am thinking of by 5 and subtract 3 the answer is 72. What number am I thinking of?

4 The perimeter of a rectangle is 24 cm. The length is 3 cm longer than the width. Work out the length and width.

5 A bus can carry 81 passengers. It arrives empty at the first stop where a certain number get on. At the second stop the number getting on is 3 more than at the first stop. Eight get off. There is now room for 44 more passengers. How many passengers got on at the first stop?

6 William has a number of 20 p, 10 p and 5 p coins in his pocket. He has twice as many 10 p coins as 20 p coins and seven 5 p coins. He has £2.75. How many 10 p coins has he?

7 The perimeter of this quadrilateral is 55.
Find x.

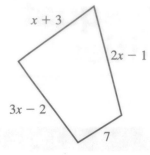

8 ABCD is a rectangle.
The perimeter is 36 cm.
Find the area.

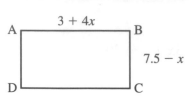

9 ABC is an equilateral triangle.
Find the length of a side.

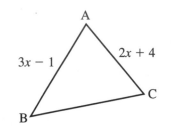

Exercise 2.3   Links: (*2H*) 2H

Make y the subject of the formula:

1 $y + x = 3p$ **2** $p + y = 3q$ **3** $2p + y = 4q$

4 $3p - y = x$ **5** $2p = y - x$ **6** $3q = 2x - y$

7 $2y = s$ **8** $ay = 7p$ **9** $2x + 3y = 17a$

10 $2x - 3y = 17a$ **11** $5y + 7p = 9q$ **12** $8a + 3b = 4x - 2y$

13 $\dfrac{y}{3} = 7p$ **14** $\dfrac{y}{5} = 2 - 3p$ **15** $\dfrac{y}{x} = 2 + p$

16 $\dfrac{1}{y} = 4x + 1$ **17** $2p - \dfrac{2}{y} = 0$ **18** $ay + by = 1$

Exercise 2.4   Links: (*2I, J, K*) 2I, J, K

1 Write down the inequalities shown on the number lines.

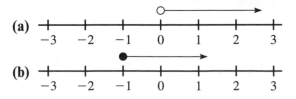

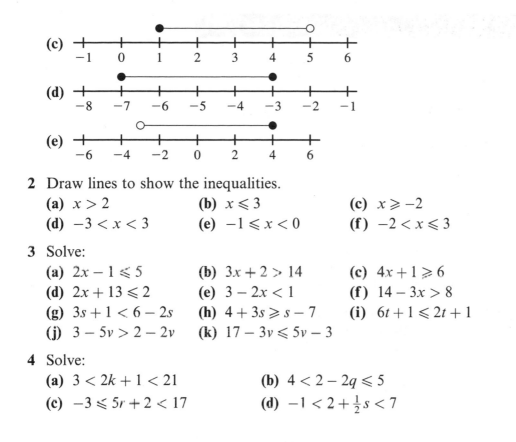

2 Draw lines to show the inequalities.
  (a) $x > 2$
  (b) $x \leqslant 3$
  (c) $x \geqslant -2$
  (d) $-3 < x < 3$
  (e) $-1 \leqslant x < 0$
  (f) $-2 < x \leqslant 3$

3 Solve:
  (a) $2x - 1 \leqslant 5$
  (b) $3x + 2 > 14$
  (c) $4x + 1 \geqslant 6$
  (d) $2x + 13 \leqslant 2$
  (e) $3 - 2x < 1$
  (f) $14 - 3x > 8$
  (g) $3s + 1 < 6 - 2s$
  (h) $4 + 3s \geqslant s - 7$
  (i) $6t + 1 \leqslant 2t + 1$
  (j) $3 - 5v > 2 - 2v$
  (k) $17 - 3v \leqslant 5v - 3$

4 Solve:
  (a) $3 < 2k + 1 < 21$
  (b) $4 < 2 - 2q \leqslant 5$
  (c) $-3 \leqslant 5r + 2 < 17$
  (d) $-1 < 2 + \frac{1}{2}s < 7$

Exercise 2.5 Links: (2L) 2L

1 List the possible integer values that satisfy:
  (a) $-3 < x < 1$
  (b) $-1 \leqslant x < 4$
  (c) $-6 < x \leqslant -3$
  (d) $-2 < 3x + 1 \leqslant 7$
  (e) $-5 \leqslant 1 - 4x \leqslant 18$

2 Write down an inequality satisfied by the integers:
  (a) 0, 1, 2
  (b) 2, 3, 4, 5, 6
  (c) $-8, -7, -6, -5$
  (d) $-1, 0, 1$
  (e) 22, 23
  (f) $-3, -2, -1$

3 Find the greatest integer value of n:
  (a) $5n + 3 < 14$
  (b) $6 - 3n \geqslant 2$
  (c) $7n + 4 < 1$
  (d) $4 - 2n > 7$
  (e) $3 + 3n < 10 - n$
  (f) $3 - 2n \geqslant 6 - n$

4 Find the smallest integer value of n:
  (a) $2n + 7 \geqslant 1$
  (b) $9 - 3n < 6$
  (c) $5n + 9 > 0$
  (d) $8 - 5n \leqslant -6$
  (e) $0 > 3 - 2n$
  (f) $5n - 5 > n + 20$

5 x is an integer, such that $-3 < x \leqslant 2$
  List all the possible values of x.   [E]

6 List all the possible integer values of x:
  (a) $5 < 1 + x < 8$
  (b) $-7 < 2x + 3 \leqslant 3$
  (c) $0 \leqslant 4 - 2x < 15$
  (d) $-8 < 2 - 3x < 1$

Exercise 2.6   Links: (2M) 2M

Solve the equations:

1 $4x - 9 = 17$ **2** $5y + 7 = 32$ **3** $7a + 23 = 9$

4 $12 + 3p = 8$ **5** $40 - 3q = 25$ **6** $6 - 7g = 34$

7 $5h + 1 = 2 - 2h$ **8** $10 + 3b = 27 + 8b$

9 $9 - 2z = 7 - 5z$ **10** $\frac{1}{2}m + 6 = 12 - m$

11 I multiply a number by 4 and add 7. The result is the same as when I take the number away from 22.

12 Coaches carry five times as many passengers as minibuses.
A fleet of 3 coaches and 11 minibuses can carry 312 passengers.
Find how many passengers each coach can carry.

13 Rearrange these formulas to make x the subject.
  (a) $p = 3 + ax$ **(b)** $4 = q - bx$
  (c) $a + 3x = b + 5x$ **(d)** $2p - 4x = 3q + 6x$

14 Solve the inequalities:
  (a) $5x - 7 < 8$ **(b)** $7x + 9 \geqslant 20$ **(c)** $10 - 4x > 9$

3 Shapes

1 Calculate the size of the lettered angles:

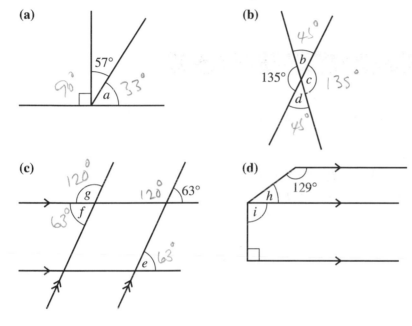

(a)

57°

90° 33°

a

(b)

45°

b

135° *c* 135°

d

45°

(c)

120°

g 120° 63°

63° *f*

e 63°

(d)

129°

h

i

2 Use only a pencil, ruler and compasses to construct an isosceles triangle with angles 30°, 30°, 120°.

3 Calculate the size of the angles marked with letters.
Write down what type each triangle is:

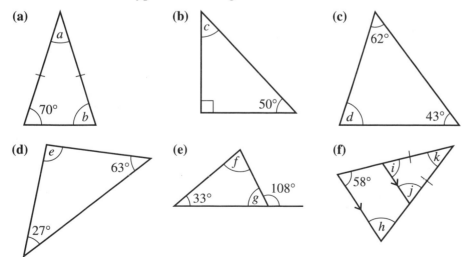

(a)

a

70° *b*

(b)

c

50°

(c)

62°

d 43°

(d)

e

63°

27°

(e)

f

33° *g* 108°

(f)

k

i

58° *j*

h

4 Construct these triangles. Use a ruler, protractor and compasses.
  (a) Triangle ABC where $AB = 10$ cm, $BC = 6$ cm, $AC = 8$ cm.
  (b) Right-angled triangle DEF where $DE = 7$ cm, $E\widehat{D}F = 90°$
      and $DF = 9$ cm. Then measure the length EF.
  (c) Triangle PQR where $QR = 5$ cm, $P\widehat{Q}R = 58°$, and
      $P\widehat{R}Q = 98°$. Then measure PR.
  (d) An isosceles triangle XYZ where $Z\widehat{Y}X = Z\widehat{X}Y = 67°$ and
      $YX = 6$ cm.

Exercise 3.2   Links: (*3C, D*) 3C, D

1 $ABCD$ is a quadrilateral. Write down the name of the
  quadrilateral if opposite sides are equal in length and the
  diagonals bisect at 90°.

2 (a) $ABCD$ is an arrowhead.   **(b)** $ABCD$ is a rectangle.
      Find: **(i)** $A\widehat{B}C$ **(ii)** $B\widehat{C}D$ Find: **(i)** $P\widehat{C}D$ **(ii)** $A\widehat{B}P$
                                                (iii) $B\widehat{C}P$ **(iv)** $A\widehat{P}D$

3 Construct a rhombus $ABCD$ where:

  $AB = 6$ cm and angle $A\widehat{B}C$ is 60°.

4 Check that Euler's Theorem works for these polyhedra:

Solid	V	F	E
Cuboid			
Pentagonal prism			
Hexagonal pyramid			
Pentagonal pyramid			
Octagonal pyramid			

Euler's Theorem says that
V + F − E = 2 where
V = the number of vertices
F = the number of faces
E = the number of edges

5 Sketch a plan, front elevation and side elevation for each of these solids.

(a) 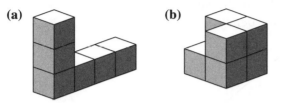 (b)

6 Sketch a net to make an octagonal pyramid.

7 Sketch the net to make this solid.

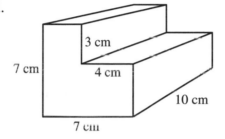

3 cm

7 cm

4 cm

10 cm

7 cm

8 Make (a) an isometric (b) an oblique projection of the solid shown by the plan and elevations.

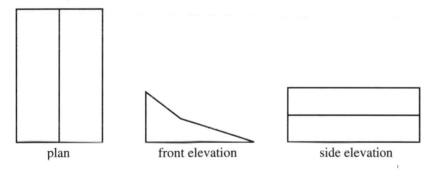

plan   front elevation   side elevation

Exercise 3.3   Links: (*3E*) 3E, F

1 (a) ABCD is a rhombus.
   Prove that ABC is congruent
   to ADC.

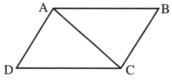

(b)

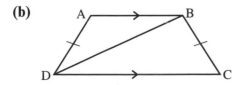

In the diagram
AD = BC
∠ADC = ∠BCD

Prove that triangles ABD and CDB are congruent.

Exercise 3.4   Links: (*3E, F*) 3E, F

1 Which of these pairs of triangles are congruent? List the vertices in corresponding order and give reasons for congruency.

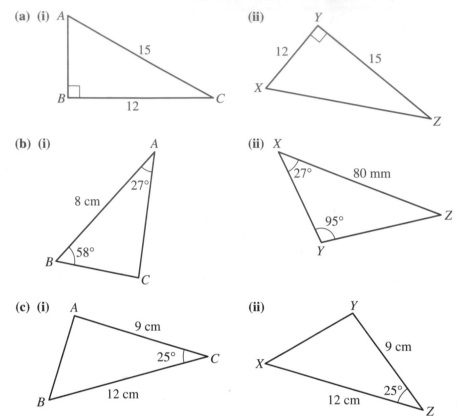

(a) (i)   **(ii)**

(b) (i)   **(ii)**

(c) (i)   **(ii)**

2 State which pairs of shapes are congruent.

3 Write down why these pairs of triangles are similar. Calculate the length of each side marked by a letter.

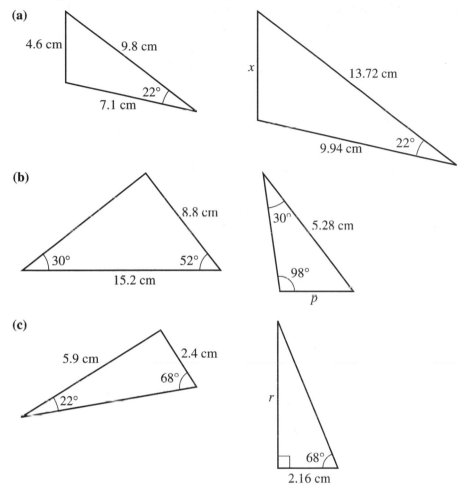

(a)

4.6 cm 9.8 cm

22°

7.1 cm

x

13.72 cm

9.94 cm 22°

(b)

8.8 cm

30° 52°

15.2 cm

30° 5.28 cm

98°

p

(c)

5.9 cm 2.4 cm

68°

22°

r

68°

2.16 cm

4 **(a)** Explain why these two triangles are similar.
(b) Calculate the lengths p and q.

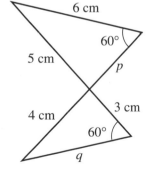

6 cm

60°

5 cm

p

4 cm

3 cm

60°

q

5 *ABCD* is an isosceles trapezium.
(a) Name the similar triangles.
(b) Explain why they are similar.
(c) Calculate the lengths
AP and *PC*.

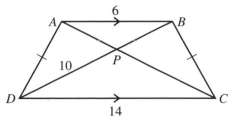

A 6 B

10 P

D 14 C

Exercise 3.5   Links: (*3G*) 3G

1 Copy these shapes. Draw in all the lines of symmetry. Write
   down the number of lines of symmetry under each diagram.

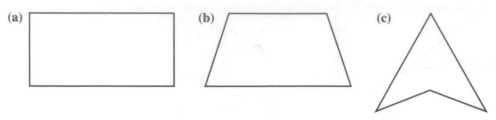

2 Write down how many planes of symmetry there are for each of
   these solids.

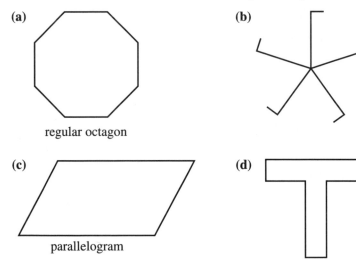

3 Write down the order of rotational symmetry of these shapes (if any).

(a)

regular octagon

(b)

(c)

parallelogram

(d)

4 Write down the shape in question **3** which has point symmetry.

5 Draw a shape that has rotational symmetry of order
   (a) 2   **(b)** 3

6 Copy and complete this diagram so that it has 3 lines of
   symmetry and rotational symmetry order 3.

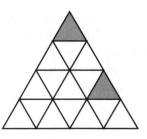

4 Collecting and presenting data

1 For each of these sets of data write down whether it is qualitative or quantitative:
   (a) the height of students at Lucea School
   (b) the students' hat size
   (c) the students' favourite food
   (d) the students' colour of eyes

2 For each of these types of data write down whether it is discrete or continuous:
   (a) the speed at which a horse can gallop
   (b) the number of apples in a bag
   (c) the number of cars in a car park
   (d) the weight of a sack of potatoes
   (e) the number of points scored by a rugby team

3 For question 1 write down whether the data can be described as either discrete or continuous.

4 For question 2 write down whether the data is qualitative or quantitative.

5 Here are the number of goals scored by a team playing soccer:

2	0	3	5	1
0	1	2	2	3
4	0	1	2	0
1	1	2	1	3
3	2	1	0	1

   (a) Draw up a frequency table for this data.
   (b) Show the data as a bar-line graph.

6 The weight of 30 boys, in kg, in a class is given below:

50.7	55.1	48.3	52.3	50.9
61.3	54.7	59.8	53.2	46.5
63.7	53.0	56.9	54.2	46.9
52.8	66.3	54.1	51.3	54.5
57.4	52.6	48.2	64.2	57.1
55.9	60.3	54.2	41.1	54.3

   (a) Draw up a table to display this data.
   (b) Display the data as a histogram.

7 Mary took a random sample of 40 people and obtained their
   shoe size. The results were:

33	45	31	39	41	39	38	44	41	39
39	40	40	42	33	43	43	45	40	45
46	34	39	37	46	38	41	36	41	39
42	41	43	41	47	41	39	42	42	40

   (a) Present this information in a stem and leaf diagram.
   (b) Use your stem and leaf diagram to work out the median of
   these 40 shoe sizes.

Exercise 4.2   Links: (*4C, D*) 4C, D

1 Write down, with reasons, whether the following questions
   would be suitable for use in a questionnaire. Suggest improved
   versions where necessary.
   (a) Do you like sweets?
   (b) What films do you watch?
   (c) Why do you play sport?
   (d) Don't you think everybody should go to the cinema every week?
   (e) Do you eat meat?

2 Design a questionnaire to investigate whether the council should
   provide more leisure facilities.

3 Design a questionnaire to investigate whether hamburgers are
   more popular with people under 16.

4 John is investigating which football teams are most popular with
   students in his school. He decided to ask 12 of his friends in the
   school football team.
   (a) Explain why this method is not reliable.
   (b) Suggest how John could make it more reliable and design a
   suitable questionnaire.

5 There are 1800 pupils at Shimpling High School. The table shows
   how these pupils are distributed by year group and gender.

Year group	Number of boys	Number of girls
9	180	195
10	189	218
11	169	191
12	162	184
13	150	162

   Philip is conducting a survey about the pupils' favourite
   television programmes. He decides to use a stratified random
   sample of 250 pupils according to year group and gender.
   (a) How many year 12 boys should be in his sample?
   (b) How many year 10 girls should be in his sample?

6 Explain how to make a selective sample of 8% of the 1500 workers at a car factory.

7 There are 1800 pupils at Irving Academy. One pupil, Sharon, wishes to take a random sample of 90 pupils for her maths project. Describe at least three different ways Sharon could take such a sample.

Exercise 4.3   Links: (*4E, F*) 4E, F

1 The frequency distribution gives the weekly sale of dresses in Bettie's Boutique for one year.

Number of dresses sold	20–29	30–39	40–49	50–59	60–69	70–79	80–89
Number of weeks	1	4	19	13	8	5	2

Draw up a frequency polygon for this data.

2 The table shows the marks obtained by 250 pupils in a test.

Mark %	Frequency
0–10	1
11–20	3
21–30	6
31–40	12
41–50	47
51–60	62
61–70	79
71–80	25
81–90	13
91–100	2

(a) Draw a cumulative frequency diagram.
(b) What percentage of pupils scored more than 64 marks?
(c) What percentage of pupils scored between 32 and 64 marks?
(d) What mark was exceeded by 75% of the pupils?

3 The table shows the age distribution of an island's population.

Age	Frequency (10 000)
0–10	38
11–20	67
21–30	102
31–40	87
41–50	48
51–60	25
61–70	12
71–80	7
81–90	4

(a) Draw a cumulative frequency diagram.
(b) Use your diagram to estimate:
  (i) the percentage of the population 16 years of age or younger
  (ii) the percentage of the population between 16 and 65 years of age

1 The marks obtained by 15 pupils in two maths tests are shown in the table.

Test 1	94	90	52	72	60	75	63	70	40	81	59	48	66	78	75
Test 2	80	86	40	74	54	70	63	65	44	78	62	46	62	74	80

(a) Draw a scatter diagram.
(b) Draw on the line of best fit.
(c) Comment on the type of correlation.
(d) Use your line of best fit to estimate a mark for Test 1 for a pupil who scores 58 on Test 2.
(e) Use your line of best fit to estimate a mark for Test 2 for a pupil who scores 32 on Test 1.

2 The table shows the minimum temperature in January and the height above sea level for ten major towns.

Height above sea level (m)	200	500	700	850	1500	2000	3800	4000	4500	4800
Jan minimum temp (°C)	4	7	5	−1	−4	−3	−10	−8	−12	−18

(a) Draw a scatter diagram for this data.
(b) Comment on the type of correlation.
(c) Estimate the minimum temperature in January for a major town whose height above sea level is 1200 m.

3 Ten pupils took two examination papers in Science. Their marks, out of 100, were:

Paper 1	88	48	80	96	60	50	20	74	78	68
Paper 2	86	56	76	84	64	60	50	70	80	74

(a) Draw a scatter diagram for these marks.
(b) Draw on the line of best fit.
(c) Comment on the type of correlation.
(d) Use your line of best fit to estimate the mark given for paper 2 to a pupil who scores 64 on paper 1.
(e) Use your line of best fit to estimate the mark given for paper 1 to a pupil who scores 78 on paper 2.

5 Using basic number skills

1 Work out:
   (a) 10% of £15   (b) 12% of £18
   (c) $17\frac{1}{2}$% of £50   (d) $17\frac{1}{2}$% of £185
   (e) 5% of 30 metres   (f) 24% of 45 centimetres
   (g) $37\frac{1}{2}$% of 150   (h) $66\frac{2}{3}$% of 297

2 Express the first quantity as a percentage of the second:
   (a) £35, £50   (b) £45, £60
   (c) 20 p, 30 p   (d) 24 m, 40 m
   (e) £17, £20   (f) £12, £8
   (g) 5 m, 6 m   (h) 5 cm, 4 cm

3 (a) Increase £50 by 5%.   (b) Increase £120 by $17\frac{1}{2}$%.
   (c) Decrease £30 by 12%.   (d) Decrease £135 by $12\frac{1}{2}$%.
   (e) Increase 56 cm by 8%.   (f) Decrease 1.2 m by 15%.
   (g) Increase £3200 by 6%.   (h) Decrease £2500 by $33\frac{1}{3}$%.

4 Gareth weighs 12 stone. He needs to lose 2 stone. What percentage of his weight is this?

5 Susan weighs 85 kg before going on a diet. She lost 6% of her original weight. What weight did she end up at?

6 Shamus buys a car for £5000. He sells it a year later for £4500. By what percentage has the car depreciated?

7 By what number do you multiply to:
   (a) increase a quantity by 12% (b) decrease a quantity by 15%
   (c) increase a quantity by 8% (d) decrease a quantity by 6%
   (e) add VAT at $17\frac{1}{2}$%   (f) give a sale price with 20% off

8 Calculate the percentage increase or decrease in these cases:
   (a) £25 to £30   (b) £2 to £2.50
   (c) $3000 to $5000   (d) 62 kg to 54 kg
   (e) £56 to £64   (f) £84 to £72

9 Rodney Trott buys 100 mobile phones for £1000. He sells them all for £7.99 each. Work out his percentage loss.

10 Tricky Dicky bought 100 mobile phones for £500. He sold them at £8.99 each. Work out his percentage profit.

11 In 1998 Roy bought a brand new car for £15 000. It is going to lose 15% of its value in the first year and 10% in every year after that. Work out the value of the car after 5 years.

12 In 1993 Rashmi bought a house for £80 000. It lost 2% of its value during the first year and then gained 3% of its value over each of the next 2 years. At the end of the fourth year the house was worth £90 000. What was the percentage increase or decrease in value of Rashmi's house during the fourth year?

13 Work out the original price in the following cases:

	New price	Percentage change	Original price
(a)	£150	10% increase	
(b)	£150	10% decrease	
(c)	£5	30% decrease	
(d)	£3500	15% increase	
(e)	£2000	2% increase	
(f)	£19.50	5% decrease	
(g)	£13.25	5% increase	
(h)	£2.99	$17\frac{1}{2}$% increase	

14 Abdul buys a TV set in a sale. The TV set cost Abdul £350 after a discount of 20%. How much money did Abdul save by buying in the sale?

15 Cecile buys a Chinese silk carpet for £500 after VAT at $17\frac{1}{2}$% has been added. What was the price before VAT was added?

16 Simon bought a house for £150,000 in 2001. A year later it had increased in value by 10%. The year after, it decreased in value by 5%. What is the house worth in 2003?

17 Ruth bought a car for £5000 in 1960. In 1970 the car was worth 10% of its value in 1960. In 2000 the car was worth 50% more than it was worth in 1970. What was the car worth in 2000?

Exercise 5.2   Links: (5*H*) 5H

1 Cheryl invests £200 at 4% compound interest for 3 years. How much money is in the account at the end of this time?

2 Work out the total amount of money when the following amounts of money are invested at compound interest:
   (a) £200 for 2 years at 5%
   (b) £1000 for 3 years at $4\frac{1}{2}$%
   (c) £250 for 4 years at 3.6%
   (d) £10 000 for 5 years at 6.1%

3 Piara invested £1000 at $5\frac{1}{2}\%$ compound interest for 5 years.
   (a) How much money is in the account at the end of 5 years?
   (b) What is the difference in the amount of interest if Piara had used simple interest and not compound interest?

4 Davindar invests money at 8% compound interest. How many years must the money be invested if it is to double in value?

5 Rachael borrows £5000 from the bank at a rate of 10% interest. She pays it back at the rate of £1000 per annum. Draw this table, extend it and use it to work out how many years it takes Rachael to repay the loan.

Year	Amount owed at start of year	Working	Amount owed at end of year
1	£5000	10% of £5000 = £500	£5000 + 500 − 1000 = £4500
2	£4500		

Exercise 5.3   Links: (5*H* – *L*) 5H – L

1 Bill gets in his car to travel from London to Edinburgh. The distance is 400 miles and it takes Bill 8 hours to get there. What is the average speed for the whole journey?

2 Nik travels 80 miles by train in 1 hour 5 minutes. Work out the average speed in miles per hour.

3 Natalie travels 85 miles from her home to London. The first 5 miles take 15 minutes, the next 70 miles take 65 minutes and the last 10 miles take 55 minutes.
   (a) Work out her average speed for each of the 3 stages of the journey.
   (b) Calculate the average speed for the whole journey.

4 Keith and David are taking part in a 20 mile sponsored walk. Keith takes 6 hours 20 minutes and David takes 40 minutes less. Work out their average speeds.

5 Daniel's car has a trip computer. It tells him how many miles per gallon of petrol he is using. He is travelling from his home in the middle of Manchester to see his mother who lives in the middle of Bristol. The total distance is 165 miles. The table shows how many miles per gallon the car uses at each stage of the journey. Work out the average number of miles per gallon the car uses on the whole journey.

Miles travelled	Miles per gallon
10	30
150	40
5	5

6 Kate used teletext to find the exchange rate between euros, pounds sterling and dollars. She found that:

    £1 was equivalent to $1.45 dollars
    £1 was equivalent to €1.62

Calculate the exchange rate between euros and dollars.

7 A cylindrical water tank full of water has a diameter of 0.6 metres and a height of 1.2 metres. It is emptying into a drain at a rate of 250 m*l* per minute.
   (a) How long will it take to empty the whole tank?

Once the tank is empty it starts to fill again at a rate of 300 m*l* per minute whilst still emptying at 250 m*l* per minute.
   (b) How long does it take the tank to completely fill again?

8 A water main has a diameter of 30 centimetres. The water flow through the pipe is at a rate of 5 metres per second. Calculate the amount of water that passes through the pipe in one hour.

9 The density of gold is 19 355 kg per m^3. Calculate the weight of a gold bar in the shape of a cuboid that has dimensions 20 cm by 8 cm by 25 mm.
   The gold bar is recast into cylindrical gold wire without losing any of the metal. The cylinder has a diameter of 2 mm. How long is the wire?

10 There were 4 farthings in a penny, 4 pennies in a groat and 12 pennies in a shilling. Calculate the number of:
   (a) farthings in 25 groats
   (b) groats in 20 shillings

Exercise 5.4   Links: (*5M*) 5M, N

1 Sandra shares a packet of 24 sweets with Mel in the ratio 5 : 3. How many sweets do each of them receive?

2 Richard and Ann share their household expenses in the ratio 3 : 2.
   (a) Work out what each of them pays if the electricity bill is £35.
   (b) Richard pays £12 as his share of the gas bill. What does Ann pay?
   (c) Ann pays £12 as her share of the council tax. What does Richard pay?

3 Luke and Melissa give 10% of their take home pay to charity.
The ratio of their take home pay is 5 : 4.
   (a) Luke gives £10 to charity one week. How much does Melissa give?
   (b) Melissa gives £10 to charity the next week. How much does Luke give?
   (c) One week the total they gave to charity was £10.80. Work out their take home pay that week.

4 The standard model car can travel 4.5 miles per litre of petrol.
The de luxe model car can travel 4.8 miles per litre of petrol.
   (a) Write down the ratio of their petrol consumption in a simpler form.
   (b) Both cars travel 620 miles. How much petrol does each car use?

5 Peter, Paul and Mary buy lottery tickets in the ratio 2 : 3 : 4.
They share their winnings in the same ratio.
How much does each receive if they win:
   (a) £10   **(b)** £84   **(c)** £5000   **(d)** £2 000 000?

Exercise 5.5   Links: (5N – Q) 5O – R

1 Write these numbers as powers of 10:
   (a) 1000   **(b)** 100 000   **(c)** 10   **(d)** 1 000 000
   (e) 1   **(f)** 10 000   **(g)** 100   **(h)** 100 000 000

2 Write these numbers in standard form:
   (a) 3450   **(b)** 239   **(c)** 376 000   **(d)** 2 460 000
   (e) 350   **(f)** 45 000   **(g)** 87   **(h)** 876 333

3 Write these numbers, that are written in standard form, as ordinary numbers:
   (a) 4.56×10^4 **(b)** 6.3×10^2 **(c)** 9×10^5 **(d)** 6.9×10^6
   (e) 7.2×10^1 **(f)** 8.1×10^3 **(g)** 4.53×10^7 **(h)** 5.332×10^5

4 Write these numbers as powers of 10:
   (a) 0.01   **(b)** 0.1   **(c)** 0.000 01   **(d)** 0.000 000 1
   (e) 0.001   **(f)** 0.0001   **(g)** 0.000 001   **(h)** 0.000 000 01

5 Write these numbers, that are written in standard form, as ordinary numbers:
   (a) 4.56×10^{-4} **(b)** 5.9×10^{-3} **(c)** 7×10^{-2} **(d)** 3.75×10^{-6}
   (e) 5.6×10^{-1} **(f)** 7.25×10^{-7} **(g)** 2.2×10^{-5} **(h)** 4.112×10^{-3}

6 Write these numbers in standard form:
   (a) 0.0234   **(b)** 0.000 32   **(c)** 0.675   **(d)** 0.000 003 45
   (e) 0.000 023   **(f)** 0.0893   **(g)** 0.002 895   **(h)** 0.000 000 561

7   Calculate, giving your answer in standard form:
     (a) $4.54 \times 10^4 + 3.76 \times 10^3$ **(b)** $2.225 \times 10^3 - 3.65 \times 10^2$ **(c)** $5.6 \times 10^5 + 6.95 \times 10^6$
     (d) $7.25 \times 10^2 - 9.75 \times 10^1$ **(e)** $5.76 \times 10^{-4} + 3.84 \times 10^{-3}$ **(f)** $2.65 \times 10^{-2} - 6.34 \times 10^{-3}$
     (g) $\dfrac{8.34 \times 10^3 + 5.98 \times 10^4}{6.53 \times 10^{-5}}$ **(h)** $\dfrac{6.98 \times 10^{-4} - 9.95 \times 10^{-5}}{5.75 \times 10^3}$

8   The distance of the Sun from the Earth is 9.32×10^7 miles. The distance of the Moon from the Earth is 2.25×10^5 miles.
     (a) How many times further is it to the Sun than to the Moon?

     Light travels at 186 000 miles per second.
     (b) How long does it take a ray of light to travel:
        (i) from the Sun to the Earth,
        (ii) from the Moon to the Earth?

9   The average distance between two atoms is 1.6 Angstrom units where 10^{10} Angstrom units is equal to 1 metre. Write 1.6 Angstrom units in metres in standard form.

10   A light year is the distance travelled by a ray of light in one year. The speed of light is approximately 3.0×10^5 kilometres per second. How far, in kilometres, is one light year? Give your answer in standard form.

Exercise 5.6   Links: (5*R*) 5S

1   Fiona buys a new flute. It costs £325. She pays a deposit of 15% and the remaining cost in 12 equal monthly instalments. Work out how much Fiona should pay each month.

2   David buys a hi-fi costing £1000 using a credit card. The credit card company charges interest at 1.95% each month. David pays £100 each month to the credit card company. How many months will it take David to pay for the hi-fi? [Hint: set your working out in a table. One has been started for you.]

Month	Working	Amount owed
1	$1000 + 19.5 - 100$	£919.50
2	$919.5 +$	
3		

3   Josie bought a new car. It cost £12 000. The car depreciated by 15% in the first year and 10% a year in subsequent years. What would be the value of the car after 3 years?

4   All the rain falling on the roof of a shed is collected in a cylindrical barrel of radius 30 cm and height 1.2 m. One day when the barrel is empty 2 cm of rain falls on to the roof. If the roof is a rectangle of length 3 m and width 1.8 m work out the height of water in the barrel.

5   Work out $\dfrac{5.46 \times 10^6 + 3.23 \times 10^4}{7.23 \times 10^{-5}}$

   Give your answer in standard index form.

6 Transformations and loci

1 Write down the three-figure bearing for each of these directions.

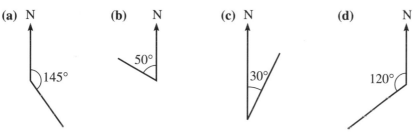

2 Write down the three-figure bearing for each of these directions:
(a) south   (b) west   (c) east   (d) south-west
(e) north-east (f) north-west (g) north   (h) south-east

3 Measure and write down the bearings of B from A in each of
these diagrams.

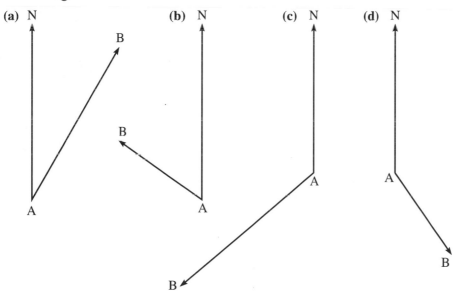

4 For each of the bearings in question 3, work out the bearing of
A from B.

5 Draw diagrams to show the following bearings.
(a) Y on a bearing of 060° from X
(b) P on a bearing of 145° from Q
(c) R on a bearing of 310° from S
(d) G on a bearing of 220° from H

Exercise 6.2   **Links:** *(6C – F)* 6C – F

1 Copy this grid into your exercise book.
   Translate the shape T using the following vectors.

   (a) $\begin{pmatrix} 5 \\ 2 \end{pmatrix}$; call it A.

   (b) $\begin{pmatrix} 5 \\ -2 \end{pmatrix}$; call it B.

   (c) $\begin{pmatrix} -4 \\ -2 \end{pmatrix}$; call it C.

   (d) $\begin{pmatrix} -1 \\ 1 \end{pmatrix}$; call it D.

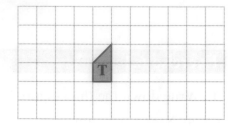

2 The shape S has been moved into four different positions
   A, B, C, D.
   Write down the vectors for each of the four translations.

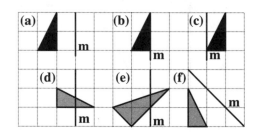

3 Copy this diagram into your exercise book.
   For each shape reflect it in the mirror line **m**.

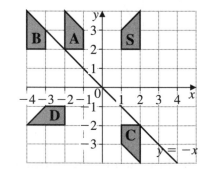

4 Shape S has been reflected four times.
   (a) Describe fully each of these reflections that takes S to
   position A, B, C, D.
   (b) Write down the rotation that moves shape A on to
   shape C.

5 Copy this diagram into your exercise book.
   Rotate each triangle about the point P marked
   with a cross by the angle and direction given.
   (a) $+90°$ **(b)** $-270°$
   (c) $-180°$ **(d)** $-90°$
   (e) $+180°$ **(f)** $+45°$

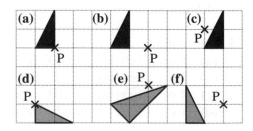

6 Shape S has been rotated four times.
   (a) Describe fully each of these rotations that takes S to
        positions A, B, C, D.
   (b) Write down the rotation that moves shape C on to
        shape A.

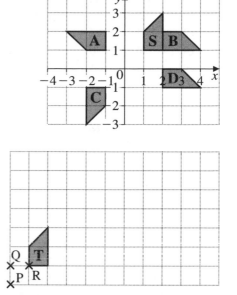

7 Copy this diagram into your exercise book.
   Enlarge shape T by the following:
   (a) scale factor 2 from point P
   (b) scale factor 3 from point Q
   (c) scale factor 2.5 from point R
   (d) scale factor $\frac{1}{2}$ from point R
   (e) scale factor $-\frac{1}{2}$ from point R

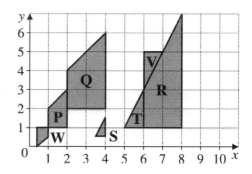

8 Describe the enlargement that moves:
   (a) shape P on to shape Q
   (b) shape T on to shape R
   (c) shape T on to shape S
   (d) shape R on to shape T
   (e) shape T on to shape V
   (f) shape P on to shape W
   (g) shape W on to shape P

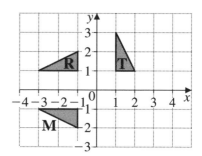

Exercise 6.3   Links: (6G) 6G

1 (a) Triangle T is transformed on to triangle R by a single
        transformation. Write down in full this single
        transformation.
   (b) Triangle R is transformed on to triangle M by a single
        transformation. Write down in full this single
        transformation.
   (c) Write down in full the single transformation that will
        move triangle M back on to triangle T.

2 On a coordinate grid with both x- and y-axes drawn from -4 to
   $+4$ plot the points $(1, 1)$, $(1, 3)$, $(2, 2)$ and $(2, 1)$. Label this shape A.
   (a) Transform the shape A by reflecting it in the x-axis. Label
        the shape B.
   (b) Rotate shape B by rotating it $-90°$ centre $(0, 0)$. Label this
        shape C.
   (c) Write down the single transformation that will move shape
        C back on to shape A.

3   On a coordinate grid with both *x*- and *y*-axes drawn from −4 to +4 plot the points (1, 1), (1, 3), (2, 2) and (2, 1). Label this shape A.
   (a) Transform the shape A by reflecting it in the *y*-axis. Label the shape B.
   (b) Rotate shape B by rotating it 90° clockwise centre (0, 0). Label this shape C.
   (c) Write down the single transformation that will move shape C back on to shape A.

4   On a coordinate grid with both *x*- and *y*-axes drawn from −6 to +6 plot the points (1, 1), (1, 3), (2, 2) and (2, 1). Label this shape A.
   (a) Transform the shape A by enlarging it scale factor 2 centre (0, 0). Label the shape B.
   (b) Rotate shape B by rotating it −90° centre (0, 0). Label this shape C.
   (c) Reflect shape C by reflecting it in the *y*-axis. Label this shape D.
   (d) Write down the single transformation that will move shape D back on to shape B.

5   On a coordinate grid with both *x*- and *y*-axes drawn from −4 to +4 plot the points (1, 1), (1, 3), (2, 2) and (2, 1). Label this shape A.
   (a) Rotate shape A by rotating it +90° centre (0, 0). Label this shape B.
   (b) Transform the shape B by reflecting it in the *x*-axis. Label the shape C.
   (c) Reflect shape C in the line *y* = *x*. Label the reflected shape D.
   (d) Transform shape D by rotating it 180° centre (−1, −1). Label the new shape E.
   (e) Write down the single transformation that will move shape E back on to shape A.

Exercise 6.4   Links: (*6H, I, J*) 6H, I, J

1   Use squared paper to show how these shapes tessellate.

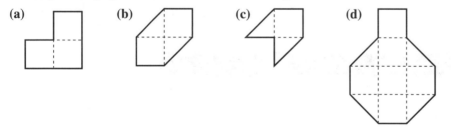

(a)   **(b)**   **(c)**   **(d)**

2   A regular octagon will not tessellate on its own. Show how you can tessellate an octagon that is not regular.

3   Albert makes a model of a fire engine. The scale of the model is 1 : 20. The height of the real fire engine is 4.2 metres.
   (a) Calculate the height of the model fire engine.

   The length of the model fire engine is 54 cm.
   (b) Calculate the length of the real fire engine.

   The ladder makes an angle of 15° with the top of the model fire engine.
   (c) Work out the angle the ladder makes with the top of the real fire engine.

4 On the Pathfinder map she was using Gill measured the distance she still had to walk as 14.5 cm. The map was drawn on a scale of 1 : 25 000. Calculate the actual distance Gill still had to walk.

5 Caesar was driving along Ermin Street and measured the distance he travelled in a straight line as 12 miles. What distance would that be on a map with a scale of 1 : 50 000?

6 Sue submitted plans for an extension to her house. The scale of the plans was 1 : 50. The length of the extension on the plan was 5.3 cm.
(a) What was the actual length of the extension?

The width of the actual extension is 3.1 metres.
(b) What measurement would the width be on the plan?

7 Marlborough is 16 miles from Swindon on a bearing of 170°. Chippenham is 20 miles from Marlborough on a bearing of 280°. Using a scale of 1 cm represents 2 miles make a scale drawing of these places.
(a) Find the bearing of Swindon from Chippenham.
(b) Calculate the distance of Swindon from Chippenham.

8 Construct the locus of the following points:
(a) 3 cm from the point Q
(b) equidistant from X and Y where XY = 5 cm
(c) equidistant from the lines AB and BC where angle ABC = 60°
(d) 2 cm from the straight line PQ where PQ = 7.5 cm

9 A goat is tethered by a 10-metre-long chain in the middle of a large field. Draw, using a scale of 1 cm to represent 4 metres, the locus of the area that the goat can graze in if the chain is attached:
(a) to a tree   **(b)** to a bar that is 20 metres long

10 Ermintrude the cow is attached by a 15 metre long chain to a bar that runs along the long side of a barn that is located in the middle of a large field as shown in the diagram. Using a scale of 1 cm to represent 2 metres draw the locus of the area in which she can graze.

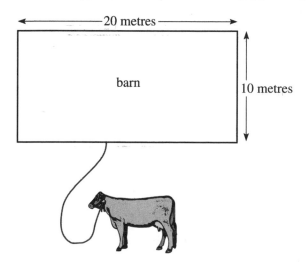

Exercise 6.5   Links: 6K

1 Using only a ruler, compass and pencil, construct triangles:
   (a) ABC so that AB = 12 cm, BC = 8 cm, AC = 6 cm
   (b) PQR so that PQ = 6 cm, QR = 6 cm, angle Q = 60°
   (c) XYZ so that XY = 8 cm, YZ = 6 cm, angle Y = 30°
   (Hint: 30° = half of 60°, so bisect 60°)

2 Draw a line AB 8.4 cm long. Draw the perpendicular bisector of AB.

3 Draw a triangle ABC so that AB = 8 cm, BC = 8 cm, AC = 10 cm.
   (a) Draw the perpendicular bisectors of AB, BC and AC. The three should meet at one point.
   (b) Place your compass point at this point and draw a circle that passes through points A,
       B and C. This is called the circumscribed circle of triangle ABC.

4 Draw a triangle PQR so that PQ = 8.5 cm, QR = 6.5 cm, PR = 9.5 cm.
   (a) Draw the angle bisectors of ∠P, ∠Q and ∠R. The three angle bisectors should meet at
       one point.
   (b) Place your compass point at this point and draw a circle that touches PQ, QR and PR.
       This is called the inscribed circle of triangle PQR.

Exercise 6.6   Links: (6J) 6K

1 A ship sails on a bearing of 210°. It then turns around and sails back along the same path.
   Work out the new bearing.

2 A is 8 km due north of B. A ship leaves A and travels on a bearing of 120°. Another ship
   leaves B and travels on a bearing of 068°. Using a scale of 1 cm to represent 1 km draw a
   scale drawing and use it to find how far from A the ships paths cross.

3 Copy this diagram into your exercise book.
   (a) Reflect the triangle T in the line $y = x$.
       Label the shape R.
   (b) Rotate the triangle T centre (0, 0) 180°.
       Label the shape S.
   (c) Describe fully the transformation that will
       move the shape S on to shape R.

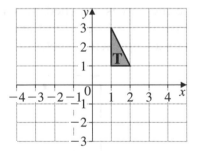

4 Mark two points P and Q, 5 cm apart. Shade in the locus of the points that are less than
   3.5 cm from P and nearer to Q than they are to P.

5 Construct an accurate drawing of a triangle *ABC* so that the lengths of the sides are
   AB = 10 cm, *BC* = 8 cm and *CA* = 6 cm. Find the point that is equidistant from *A*, *B* and *C*.

6 Construct an accurate drawing of a triangle *ABC* so that the lengths of the sides are
   AB = 10 cm, *BC* = 8 cm and *CA* = 6 cm. Find the point whose perpendicular distance from
   each side of the triangle is the same.

7 Lines, simultaneous equations and regions

1 On separate diagrams draw lines with these equations:
   (a) $y = x + 1$ (b) $y = 2x + 2$ (c) $y = 3x - 1$
   (d) $y = 4 - x$ (e) $y = 3 - 2x$ (f) $y = \frac{1}{4}x$

2 For each line in question **1**, write the equation of two lines which are parallel.

3 Write down the intercepts on the y-axis of the lines with these equations:
   (a) $y = 2x + 3$ (b) $y = 4 - 2x$ (c) $y = 3x + 7$
   (d) $y = 2 + \frac{1}{2}x$ (e) $y = -\frac{1}{2}x - 5$ (f) $y = 3\frac{1}{2}x - 1$

4 A line parallel to $y = 2x + 3$ meets the y-axis at $(0, -2)$. Write down the equation of the line.

5 A line with equation $y = 3x + a$ crosses the y-axis at $(0, 3)$. Write the equation of the line.

6 A line with equation $y = c - 2x$ passes through the point $(2, 3)$. Write down the equation of the line.

7 A line with equation $y = ax + 3$ passes through the point $(3, 0)$. Work out the equation of the line.

8 Find the intercept on the x-axis of the lines:
   (a) $y = 2x - 4$ (b) $y = 6 - 3x$
   (c) $y = \frac{1}{2}x + 7$ (d) $y = 14 - \frac{1}{3}x$

9 Write down the equation of the line with gradient 4 and y-intercept at $(0, 3)$.

10 Write down the equation of the line with gradient $-\frac{1}{2}$ and y-intercept at $(0, -4)$.

11 Write down the gradient and y-intercept of the lines:
   (a) $2x + y = 7$ (b) $y - 3x + 2 = 0$ (c) $2x + 3y = 5$
   (d) $2x - 3y + 1 = 0$ (e) $14 - 2x - 7y = 0$ (f) $-7 = 3x - 2y$

12 The gradient of a line is 5. It passes through the point $(1, 2)$. Work out the equation of the line.

13 The point $(3, -2)$ lies on a line with gradient -3. Work out the equation of the line.

14 Find the equation of the line that passes through the points $(2, 7)$ and $(14, 31)$.

15 Find the equation of the line which passes through the points $(-3, -1)$ and $(0, 11)$.

16 Find the equation of the line which passes through $(-3, 6)$ and $(3, 3)$.

17 Pick out the pairs of perpendicular lines from this list of equations.
   (a) $y = 2x + 7$ **(b)** $y = \frac{1}{2}x - 3$ **(c)** $y = x + 4$
   (d) $y = 3 - \frac{1}{2}x$ **(e)** $2y = 4 - 2x$ **(f)** $y + 2x = 3$

18 Find an equation of the line perpendicular to $y = 3x - 2$ which goes through:
   (a) $(0, 0)$ **(b)** $(3, 3)$

19 Find an equation of the line perpendicular to $2y = x + 1$ which goes through:
   (a) $(0, 0)$ **(b)** $(5, 2)$

19 Find an equation of the line perpendicular to $3y + 2x = 7$ which goes through:
   (a) $(0, 0)$ **(b)** $(2, -4)$

Exercise 7.2   Links: (*7D, E, F*) 7D, E, F

1 The points $(a, 2)$; $(3, b)$; $(-5, c)$; $(d, 2d)$ lie on the line
   $4x + y - 3 = 0$. Find a, b, c, d.

2 The points $(-1, a)$; $(b, -3)$; $(2c, 3c)$ lie on the line
   $2x - 3y + 6 = 0$. Find a, b, c.

3 Use the graph to solve:

   $x - y = 1$

   $\frac{1}{2}x - y = -1$

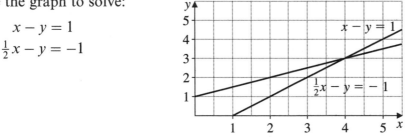

4 Draw appropriate straight lines to solve these pairs of simultaneous equations.
   (a) $2x + y = 3$ **(b)** $5x + 2y = 12$ **(c)** $4x + y = 14$ **(d)** $2x - 3y = 7$
       $x + 3y = 4$ $3x - y = 5$ $x - 2y = -1$ $x + y = 1$

5 Solve the following simultaneous equations.
   (a) $3x + 4y = 10$ **(b)** $5x + 2y = 22$ **(c)** $2x + 3y = 4$ **(d)** $x + 5y = 8$
       $2x + 4y = 8$ $3x + 2y = 10$ $2x - y = -4$ $4x + 5y = 2$
   (e) $7x + 3y = 9$ **(f)** $6x - 3y = 21$ **(g)** $2x + 5y = 13$ **(h)** $3x - 5y = 9$
       $2x - 3y = 18$ $6x + y = 25$ $x + 3y = 2$ $x + y = 2$
   (i) $4x - 3y = -17$ **(j)** $3x + 2y = 15$
       $2x + y = 3$ $2x - 3y = 9$

Exercise 7.3   Links: 7G

Set up equations and solve.

1 Seats at the cinema are £4 and £5. There were 223 customers who paid a total of £963. How many bought £4 seats?

2 Boxes of chocolates weighing 200 g and 500 g were packed into a crate. The total weight of the boxes of chocolates was 16 kg and there were 50 boxes in total. How many of each were there?

3 A string of decorative lights consists of 15 watt and 20 watt bulbs. The total wattage is 3000 watts (3 kilowatts) and there are 161 bulbs in total. Work out how many of each there are.

4 A father is x years old and his daughter y years old. Three times the difference in their ages is 57. In 2 years time the father will be twice as old as his daughter. How old are they now?

5 In a game 7 reds and 4 blues scores 29 points. 5 reds and 7 blues also scores 29 points. How many points are scored for a red?

6 At a flower shop 23 roses and 21 carnations cost £7.00; 15 roses and 30 carnations cost £7.50. Work out the cost of a single carnation.

7 The line $px - gy = 17$ goes through $(3, -4)$ and $(7, 2)$. Find p and q.

Exercise 7.4   Links: (7H) 7H

In questions **1–12**, draw diagrams to show the regions which satisfy the inequalities.

1 $x < 5$ **2** $y \geqslant -3$ **3** $-1 < x \leqslant 2$ **4** $1\frac{1}{2} \leqslant y < 2\frac{1}{2}$

5 $x + y < 3$ **6** $x - y > 2$ **7** $3x + y > 0$ **8** $y \leqslant 4x$

9 $x > 2 - 3y$ **10** $2x + 3y \leqslant 6$ **11** $4y - x > 0$ **12** $-2 < 2x + 1 < 5$

13 On a graph shade the region for which:

$$x + 2y \leqslant 6, \quad 0 \leqslant x \leqslant 4 \quad \text{and} \quad y \geqslant 0.$$ [E]

14 Draw a diagram to show the region which satisfies:

$$y < x + 1; \quad y + 2 > 2x; \quad 3y + 2x > 6.$$

15 Sketch the region defined by the inequalities:

$$2x = 3y < 12 \, ; \, y < 3x \, ; \, x > 2y.$$

16 Find the inequalities which define the shaded area:

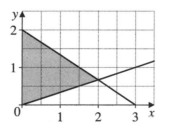

8 Pythagoras' theorem

1 Find the hypotenuse in these right-angled triangles:

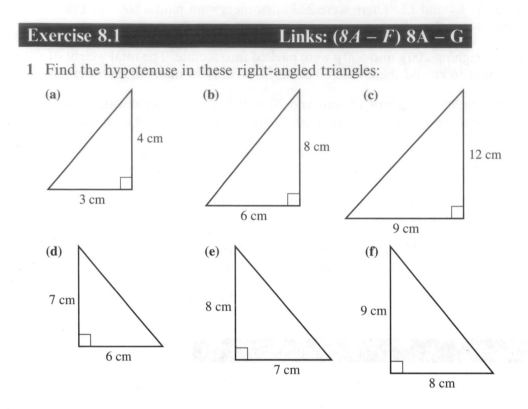

(a) 4 cm, 3 cm

(b) 8 cm, 6 cm

(c) 12 cm, 9 cm

(d) 7 cm, 6 cm

(e) 8 cm, 7 cm

(f) 9 cm, 8 cm

2 Calculate the unmarked side in these right-angled triangles:

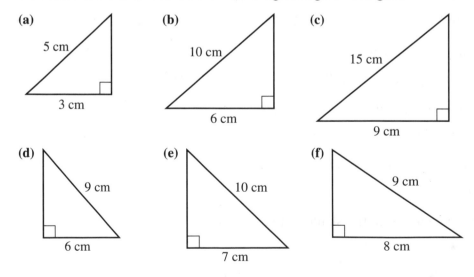

(a) 5 cm, 3 cm

(b) 10 cm, 6 cm

(c) 15 cm, 9 cm

(d) 9 cm, 6 cm

(e) 10 cm, 7 cm

(f) 9 cm, 8 cm

3 A ladder is resting against the wall of a house. The foot of the ladder is 3 m from the base of the wall and the top of the ladder is 4 m from the base of the wall. How long is the ladder?

4 Calculate the span of the roof truss shown in the diagram.

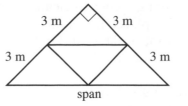

5 Calculate the lengths of the sides marked by a letter.

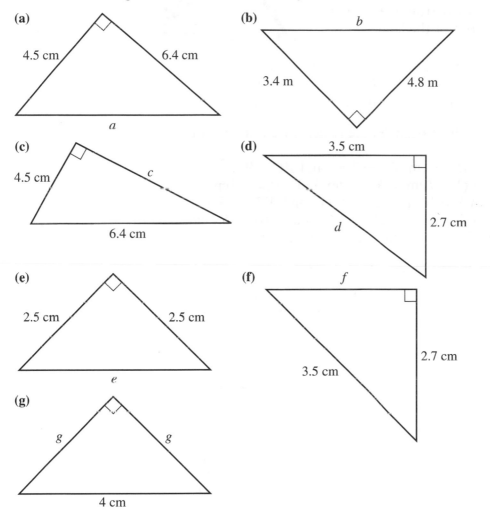

(a)

4.5 cm 6.4 cm *a*

(b)

b 3.4 m 4.8 m

(c)

4.5 cm *c* 6.4 cm

(d)

3.5 cm 2.7 cm *d*

(e)

2.5 cm 2.5 cm *e*

(f)

f 2.7 cm 3.5 cm

(g)

g *g* 4 cm

6 Susie is flying her kite on a horizontal playing field. The string is taut and the kite is 100 m above the ground. The kite is 300 m from Susie in a horizontal direction.
How long is the kite string?

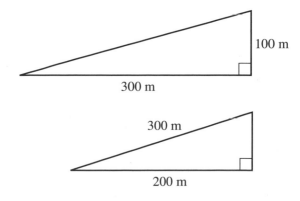

100 m

300 m

7 Susie's friend Sophie is flying her kite on the horizontal playing field. The 300 m length of string on her kite is taut and the kite is 200 m away from Sophie in a horizontal direction.
How far from the ground vertically is the kite?

300 m

200 m

8 Keith used his 6-metre-long ladder to clean
his upstairs windows. He placed the ladder 2
metres away from the foot of the wall. How
far up the wall did the ladder reach?

9 Meg used her 8-metre-long ladder to
paint her upstairs windows. She placed
the top of the ladder 6 metres above the
ground. How far away from the base of
the wall was the foot of the ladder?

10 Calculate the distance between each pair of points:
   (a) (2, 3) and (5, 8)   (b) (8, 3) and (5, 0)
   (c) (−2, 3) and (5, 0)   (d) (−3, −2) and (−5, −5)

11 Work out whether these triangles are scalene, right-angled or
   obtuse-angled:
   (a) *ABC*; where $AB = 6$ cm, $BC = 8$ cm and $AC = 9$ cm
   (b) *PQR*; where $PQ = 7$ cm, $QR = 4$ cm and $PR = 4.5$ cm
   (c) *XYZ*; where $XY = 10$ cm, $YZ = 24$ cm and $XZ = 26$ cm
   (d) *DEF*; where $DE = 6.2$ cm, $EF = 4.5$ cm and $DF = 5.4$ cm
   (e) *KLM*; where $KL = 12.1$ cm, $LM = 8$ cm and $KM = 4.9$ cm

12 Calculate the length of PQ in each of these shapes.

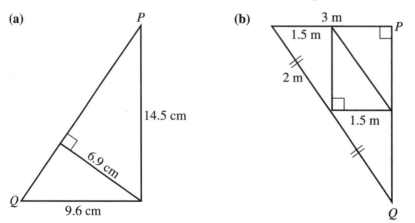

13 Write down the equations of these circles:

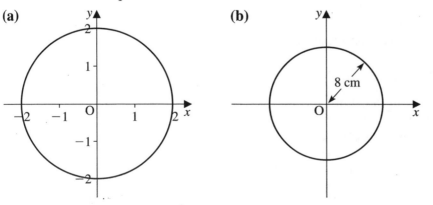

14 Find the equations of these circles:
  (a) radius 10 with centre the origin
  (b) radius 8 with centre the origin
  (c) radius 1 with centre the origin
  (d) radius 5 with centre (2, 2)
  (e) centre the origin passes through (6, 8)

Exercise 8.2   Links: (*8G*) 8H

1 Calculate the longest diagonals *AB* of these cuboids.

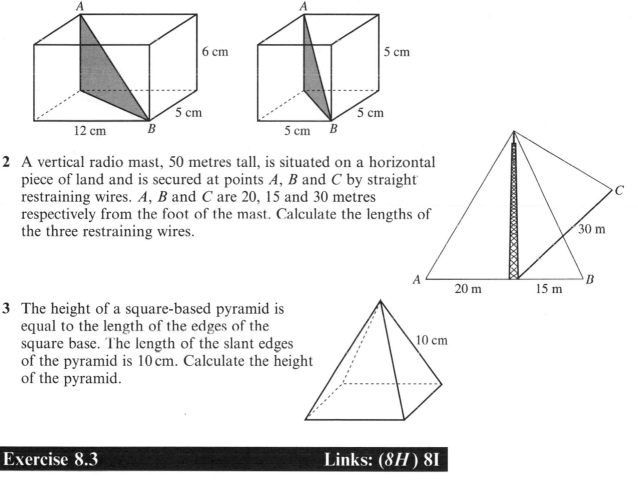

2 A vertical radio mast, 50 metres tall, is situated on a horizontal piece of land and is secured at points *A*, *B* and *C* by straight restraining wires. *A*, *B* and *C* are 20, 15 and 30 metres respectively from the foot of the mast. Calculate the lengths of the three restraining wires.

3 The height of a square-based pyramid is equal to the length of the edges of the square base. The length of the slant edges of the pyramid is 10 cm. Calculate the height of the pyramid.

Exercise 8.3   Links: (*8H*) 8I

1 Peter and Elaine sail due south from Portsmouth for 5 miles. They then sail due west for 4 miles before turning for home. How far from Portsmouth is the boat when it turns for home?

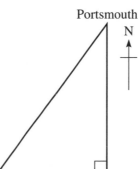

2 Isosceles triangle *ABC* has two equal
sides of length 8 cm and a base of length
6 cm. Calculate the height of the triangle.

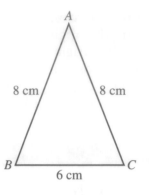

3 Isosceles triangle *PQR* has two equal sides of length 15 cm and a
height of 10 cm. Calculate the length of the base of the triangle.

4 A roof truss *XYZ* in the shape of an isosceles right-angled
triangle has a span of 20 metres. Calculate the length of one of
the slanted edges.

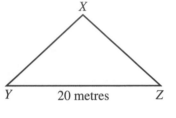

5 The diagram represents the side view of a
staircase. *PQ* represents the bannister
rail. Calculate the length of *PQ*.

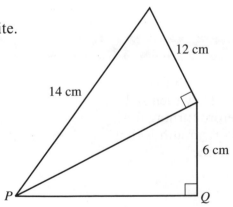

6 Work out the longest diagonal of a cube of edge 10 cm.

7 Two different right-angled triangles have lengths x, $2x$ and
$x + 5$. Find the 2 possible values of x and hence calculate the
lengths of the three sides.

8 Calculate the length of the
side *PQ* in the diagram opposite.

9 Work out the distance between the points $(-2, 5)$ and $(3, -4)$.

10 Check which of these sets of numbers is a Pythagorean triple:
    (a) 7, 24, 25   **(b)** 7, 9, 11   **(c)** 4.5, 6.0, 7.5

11 A rhombus *PQRS* has sides of length 5 cm. Calculate the length
    of the diagonal *PR* if the diagonal *QS* has length 8 cm.

12 A circular cone has a base diameter of 10 cm and a slant height
    of 15 cm. Calculate the vertical height of the cone.

13 A drinking straw of length 20 cm is put in a cylindrical glass of
    radius 3 cm and height 12 cm. Work out the least possible length
    of straw which sticks out above the rim of the glass.

14 A garage is 6 metres long, 3 metres wide and 2.5 metres high.
    Explain, showing your working how a wooden pole of length
    7 metres can be stored in the garage.

15 Calculate the diagonal length of the label of a tin of beans that
    has a radius of 3.5 cm and a height of 10 cm. [You may ignore
    any overlap.]

9 Probability

1 Twelve equal-sized balls are labelled from 1 to 12.
   The balls are placed in a bag.
   One of the balls is selected at random.
   Work out the probability that the selected ball will be labelled:

   (a) 3   **(b)** a multiple of 3
   (c) a prime number   **(d)** an even number
   (e) an odd number   **(f)** a triangular number
   (g) a square number

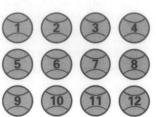

2 Fifteen chocolates are put into a bag.
   8 of the chocolates are milk, 4 of the chocolates are plain and
   the remainder are white.
   Suzanne selects a chocolate from the bag at random.
   Calculate the probability that this chocolate will be:

   (a) milk   **(b)** not milk
   (c) not white   **(d)** either white or milk

3 A selection pack contains 36 packets of crisps. In the pack there are

       12 bags of plain crisps, 8 bags of cheese and onion crisps,
        7 bags of salt and vinegar crisps, 5 bags of smoky bacon crisps
   and 4 bags of prawn cocktail crisps.

   On a dark night Savita selects a packet of crisps from the pack
   at random.
   Write down the probability that she will select:

   (a) a bag of plain crisps
   (b) a bag of crisps that are not plain
   (c) a bag of beef and onion crisps
   (d) a bag of smoky bacon crisps
   (e) a bag of crisps which is either salt and vinegar or prawn cocktail
   (f) a bag of crisps which is neither plain nor cheese and onion.

4 Kevin has a spinner in the shape of a regular pentagon, and a
   normal dice.
   The five sections of the spinner are labelled 1, 2, 3, 4, 5.
   Kevin spins the spinner once and rolls the dice once.
   He records the outcome, writing the number shown on the spinner
   first.
   This outcome is recorded as (5, 3).
   (a) Write down the full list of all the possible outcomes.

(b) Find the probability of the outcomes:
    (i) (2, 4)   **(ii)** (3, 7)   **(iii)** the two numbers add up to 10
    (iv) both of the numbers are prime
    (v) either one of the numbers is 3
    (vi) the difference between the two numbers is 1
    (vii) the second number is double the first number

5 Three coins are tossed simultaneously.
  (a) Using H and T for heads and tails, list all of
    the 8 possible outcomes.
  (b) Work out the probability that:
    (i) all 3 coins land heads
    (ii) two coins land tails
    (iii) at least one coin lands tails
    (iv) at least two of the coins land heads

6 A 'singles only' dart board has 20 equal
sectors marked from 1 to 20. It has no
bullseyes and no spaces for doubles
or trebles. An ordinary die is a cube
with its six faces marked from 1 to 6.
When Marco throws a dart at the
board he will hit one of the numbers
at random.

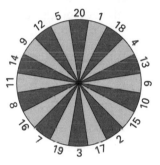

Marco rolls the die and throws a dart at the board.
The number on the upper face of the die will be used as the
x-coordinate and the score on the dartboard will be used as the
y-coordinate of a point on a grid. Calculate the probability that
when the die is rolled and the dart is thrown the point on the
grid generated will lie on the line:
  (a) $x = 4$
  (b) $y = 9$
  (c) $y = 2x$
  (d) $x = y$
  (e) $y = 3x$
  (f) $x + y = 10$
  (g) $y = 2x + 1$
  (h) $y = 20 - x$
  (i) $x - y = 3$
  (j) $x + 3y = 8$

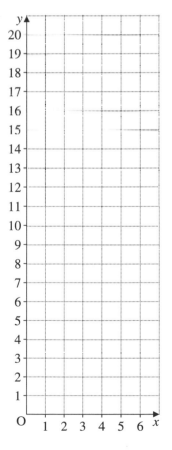

1 You are provided with the final scores from 100 first class
   football matches.

2–1	1–1	0–0	3–1	1–3	2–0	2–0	2–1	6–0	3–1
3–0	5–1	0–0	2–1	0–3	1–1	1–2	0–7	4–2	5–0
2–0	3–0	2–1	4–1	2–0	5–1	1–2	1–3	3–2	0–3
1–2	1–2	0–0	0–0	0–1	2–3	1–0	4–2	1–0	1–1
0–0	1–2	1–3	2–0	1–1	1–0	1–1	2–0	0–0	0–1
1–1	3–1	2–2	1–1	0–1	2–1	1–4	2–0	3–1	5–0
1–2	1–1	1–1	1–1	0–2	0–1	4–1	3–2	2–1	1–2
2–0	2–1	1–0	1–1	1–1	3–1	0–0	1–3	1–2	2–1
4–2	3–0	0–1	1–0	2–2	1–1	2–1	1–1	0–1	0–0
6–1	4–0	0–0	6–1	1–1	1–1	3–1	2–0	0–0	1–1

   The first score is the number of goals scored by the home team.
   The second score is the number of goals scored by the away
   team.

   A first class football match is chosen at random.
   (a) Using this evidence, and this evidence alone, make estimates
   of the probability that:
      (i) the home team will win
     (ii) the away team will win
    (iii) the game will end as a draw
    (iv) the final score will be 1–1
      (v) at least one of the teams will score no goals
    (vi) there will be a total of 3 goals in the match
   (vii) there will be at least 3 goals in the match
   (viii) there will only be a one goal difference between the two
        scores

   (b) Next month there will be exactly 250 first class football
   matches played. Using only the evidence from the data you
   have been provided with, give the best estimate for:
     (i) the number of these matches that will finish as a draw
    (ii) the number of matches in which there will be 4 or more
       goals scored

2 Shortly before an election a survey was conducted to find out
   information about people's voting intentions.
   1300 people were asked which of the four candidates they
   intended to vote for. The results of the survey were:

Candidate	Number who intend to vote
Adey	250
Burrows	450
Cresson	372
Davis	228

(a) On the day of the election a person was stopped in the street at random as they were going to vote. Work out, giving your reasons, the best estimate of the probability that this person intended to vote for:

    (i) Burrows   **(ii)** Davis

(b) In the actual election 16 500 votes will be cast. Give, with reasons, the best estimate of the number of votes cast for:

    (i) Adey   **(ii)** Cresson

Exercise 9.3   Links: (*9D, E*) 9D, E

1 Next Monday, Steve and Linda are both taking their driving test.
The probability of Steve passing is 0.6.
The probability of Linda passing is 0.7.
Work out the probability of:

(a) Steve not passing

(b) Linda not passing

(c) both Steve and Linda passing

(d) both Steve and Linda not passing

(e) Steve passing and Linda not passing

(f) Steve not passing and Linda passing

2 A bag contains 12 equal-sized coloured balls.
6 of the balls are red, 4 of the balls are blue and 2 of the balls are white.
A ball is selected at random, its colour is recorded and then the ball is put back in the bag.
A second selection is then made at random; the colour of this ball is also recorded.
Copy and complete the tree diagram.

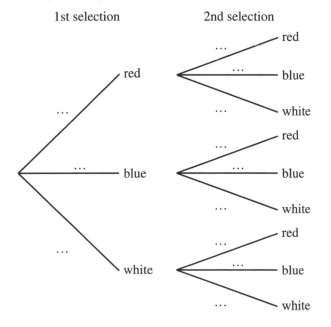

Work out the probability of the selected balls being:
   (i) both red
   (ii) both of the same colour
   (iii) one red and one blue
   (iv) of different colours
   (v) at least one blue

3 When Jaqui and Wendy play tennis the probability of
   Jaqui winning is $\frac{3}{4}$ and the probability of Wendy
   winning is $\frac{1}{4}$.
   When Jaqui and Wendy play chess the probability of
   Jaqui winning is $\frac{1}{5}$ and, like tennis, one of the two girls
   must win – assume there is no chance of a game of chess
   being a draw.

   Jaqui and Wendy play a game of tennis and a game of
   chess.
   (a) Draw a probability tree diagram for this situation.
   (b) Using your tree diagram or otherwise, work out the
       probability of:
         (i) Jaqui winning both games
        (ii) Wendy winning at least one of the games
       (iii) the girls winning one game each

4 Nikki and Ramana both try to score a goal in netball.
   The probability that Nikki will score a goal on her first try is 0.65.
   The probability that Ramana will score a goal on her first try is 0.8.
   (a) Work out the probability that Nikki and Ramana will both
       score a goal on their first tries.
   (b) Work out the probability that neither Nikki nor Ramana
       will score a goal on their first tries.   [E]

5 Just before a by-election 1500 people were asked which political
   party they intended to vote for.
   The results of the survey were:

Conservative	420
Labour	700
Lib. Dem.	300
Green Party	80

   There were no other parties standing for election.
   On the day of the election two people were stopped at random
   just before they cast their votes.
   Work out, giving your reasons, the best estimate of the
   probability of:
   (a) these two people both voting labour
   (b) these two people both voting for the same party
   (c) these two people voting for different parties

6 Here are diagrams of two spinners.

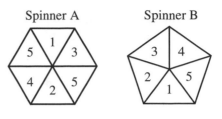

Naomi will spin each spinner once and record the number it lands on.
   (a) Work out the probability of these two numbers:
   (i) both being 5   **(ii)** both being the same
   (iii) having a sum of 4
   (b) Work out the probability of the difference between the numbers being
   (i) zero   **(ii)** 1

7 One evening, Lucy and Emma are going to play two games, darts and chess.
In darts the probability that Lucy will win is 0.7.
In chess the probability that Emma will win is 0.6.
The outcomes of the games are independent of each other.
In both games, if Lucy does not win then Emma wins. The games cannot finish as a tie.
   (a) Copy and complete the tree diagram

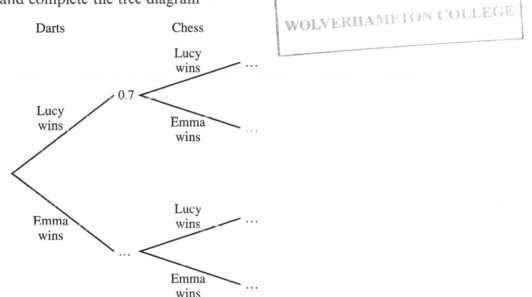

   (b) Using the tree diagram or otherwise, work out the probability that
   (i) Lucy will win both games
   (ii) Emma will win at least one game

8 Sumreen and Asif are both about to take a music examination.
The probability that Sumreen will pass is 0.8.
The probability that Asif will pass is 0.75.
Work out the probability that
   (a) they will both pass
   (b) they will both fail
   (c) at least one of them will pass

9 Anna is due to take her driving test. The probability that she will pass at the first attempt is 0.6. If she fails at the first attempt, the probability of her passing at the second, or any other, subsequent attempt is 0.8.

(a) Complete the probability tree diagram:

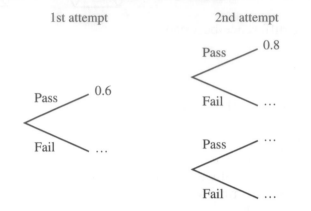

(b) Work out the probability of Anna passing the driving test:
   (i) At the second attempt
   (ii) Within, at most, two attempts
   (iii) Within, at most, three attempts

10 Jim is about to take a music test. The probability of him passing the test at the first attempt is p. If he fails, the probability of him passing on the second or any subsequent attempt is q.

Prove that the probability of Jim passing the test in no more than two attempts is

$$p + q - pq$$

10 Brackets in algebra

Evaluate and compare:

1. $(6 \times 9) + 3$ with $6 \times (9 + 3)$
2. $(5 + 3) \times (2 + 7)$ with $5 + (3 \times 2) + 7$
3. $(7 - 5) - 2$ with $7 - (5 - 2)$
4. $(8 \div 4) + (4 \div 2)$ with $8 \div (4 + 4) : 2$
5. $(7 \times 3) \times 2$ with $7 \times (3 \times 2)$
6. $(54 \div 9) \div 3$ with $54 \div (9 \div 3)$

Copy the following calculations and insert brackets so that they are correct.

7. $8 - 3 \times 2 + 4 = 6$
8. $8 - 3 \times 2 + 4 = 30$
9. $8 - 3 \times 2 + 4 = 14$
10. $8 - 3 \times 2 + 4 = -10$

Expand:

1. $3(2a + 5)$
2. $4(b - 6)$
3. $5(2 - 3x)$
4. $-2(x + 1)$
5. $-3(2 - 4y)$
6. $a(2x + 3)$
7. $ab(3y - z)$
8. $(z - 2y)x$
9. $x(3x + y)$
10. $y^2(2y + 1)$
11. $pq(q + p)$
12. $(2s + 3t)st$

Factorize completely:

13. $4x + 20$
14. $2y + 16$
15. $3a + ab$
16. $xy - 2y$
17. $3z^2 + 9$
18. $4x + 8x^2$
19. $2ax - axy$
20. $7z^2 + 21z$
21. $ay + ax$
22. $15ab - 5ac$
23. $12ab^2 - 3ab$
24. $8x^2y^2 - 2xy$

Expand and simplify:

25. $2(x + 3) + 3(2x + 1)$
26. $4(z - 3) - 3(2z + 1)$
27. $5(2y - 3) + 2(y + 1)$
28. $8(3 - 2a) - 2(a - 3)$
29. $6(3a - 4) - (a - 2)$
30. $(3y + 1) - 4(2 - 2y)$
31. $5x - 3(x - 1)$
32. $4a(b + 1) + b(a - 1)$
33. $2x(3 + 2z) - z(3x + 1)$
34. $5g(g + 3) - g(2 - g)$

Exercise 10.3   Links: *(10E, F, G)* 10E, F, G

Solve the equations:

1 $3(x+3) = 21$ **2** $4(y-5) = 24$

3 $2(5-x) = 10$ **4** $6(3x+1) = 42$

5 $-3(2-2x) = 12$ **6** $2(2x+1) - 3(x+7) = 5$

7 $4(3-2x) - 3(1-2x) = 13$ **8** $7(4x-3) + 2(5-x) = 93$

9 $3(2x+5) = 5(x-7)$ **10** $4(x+1) + 3(5x-1) = (1-x)$

11 $-4(2-3x) + 7(5x-3) = 65$ **12** $3p(p+1) - p(3p-4) = 28$

Write equations and solve these problems:

13 I add 3 to a number and multiply the result by 5. The answer is 40. What is the number?

14 I multiply a number by 5 and add 3 to it. I multiply the result of this by 2 to get a final result of 56. What is the number?

15 I take a number away from 20 and multiply the result by 7. The answer is 63. What is the number?

16 In the Premier League there are 3 points for a win and 1 point for a draw. After 23 games Manchester United have 49 points and have lost 4 matches. How many matches have they drawn?

Solve the inequalities:

17 $2(x-7) > 4$ **18** $3(2x+1) \leqslant 12$

19 $4(2-3x) \leqslant -16$ **20** $5(3x-4) > 3(x+1)$

21 $2(4-x) \geqslant x+5$ **22** $6(2x+7) < 3x-7$

23 $2(3x+1) - (5x-1) < 3(3x+6)$

24 $4(2x+3) - 5(3-2x) \geqslant 3(4-3x)$

Exercise 10.4   Links: *(10H)* 10H

Make *a* the subject of each equation:

1 $ax = ab + c$ **2** $a(x-b) = 3$

3 $3a + ab = 6$ **4** $2a + x = ab$

5 $b(a+3) = c(2a+4)$ **6** $2x(4-a) = a + 3y$

7 $\dfrac{x}{a} = 2y + z$ **8** $y = \dfrac{x}{a} + b$

Exercise 10.5   Links: (*10I*) 10I

Solve the simultaneous equations by substitution:

1 $y = 2x + 1$
    $3y + x = 24$

2 $3y - 4 = x$
    $2y + 3x = 43$

3 $a = 6 - x$
    $8 = 3x - 2a$

4 $c = 2d + 7$
    $5d = 3c - 4$

5 $k = -3j - 2$
    $2j + 5k = 3$

6 $p - -\frac{1}{2}q + 1\frac{1}{2}$
    $4p - q = 7$

7 $2p = y - 7$
    $3y - 5p - 15$

8 $3p = 8 - 2x$
    $x - 6p + 21 = 0$

In questions **9**–**12** express a in terms of t:

9 $a = 3x + 2y$ $x = 2t + 1$ $y = 1 - t$

10 $a = 2x - 5y$ $x = 1 - 3t$ $y = t + 4$

11 $a = 4x - y$ $x = 5t + 1$ $y = 10t$

12 $a = 3(x + y)$ $x = 1 - 2t$ $y = 2 - t$

Exercise 10.6   Links: (*10J*   *M*) 10J – M

Expand and simplify:

1 $(x + 3)(x + 2)$

2 $(x + 7)(x - 3)$

3 $(x - 4)(x - 5)$

4 $(x + 1)(x - 5)$

5 $(x + 7)^2$

6 $(2 - x)(x + 4)$

7 $(3 - 2x)(4 - 3x)$

8 $(5x + 2)(2 - 5x)$

9 $(x + 3)(x - 3)$

10 $(10 - 2x)(10 + 2x)$

11 $(3x + 1)(2x - 2)$

12 $(4x - 3)(2x - 7)$

13 $(3 + 5x)(2x - 1)$

14 $(7x - 3)(2 + 3x)$

15 $(2x - 3)(2x + 3)$

16 $(7 - 3x)(7 + 3x)$

Note :   There are no Practice exercises for
       Unit 11: Using and applying mathematics.

12 Estimation and approximation

Exercise 12.1   Links: (*12A, B*) 12A, B

1 Round these numbers to the nearest 10:
   (a) 379   (b) 438   (c) 94   (d) 5999

2 Round these to the nearest whole number:
   (a) 4.6   (b) 0.499   (c) 11.85   (d) 5.515

3 Rewrite these with each number rounded to the nearest whole number:
   (a) 3.8×7.4 (b) 0.51×6.98 (c) 8.6×3.9

4 54 498 people watched a football match.
   (a) Write this number to the nearest:
        (i) 10   (ii) 100   (iii) 1000   (iv) 10 000
   (b) Which figure is it most sensible to use in a newspaper headline and why?

5 Write these numbers correct to the approximation given in brackets:
   (a)   0.054926   (4 d.p.)
   (b) 458.349   (1 d.p.)
   (c)   50.0509   (3 d.p.)
   (d)   9.999   (2 d.p.)

6 Carry out the following calculations. Give your answers correct to the number of decimal places given in brackets:
   (a) 18.47×1.563 (2 d.p.)
   (b) 3.142×9.666 (3 d.p.)
   (c) 19.698×25.923 (1 d.p.)

Exercise 12.2   Links: (*12C, D*) 12C, D

1 Write these numbers correct to the approximation given in brackets:
   (a)   40.49   (3 s.f.)
   (b)   99.98   (2 s.f.)
   (c)   0.04543 (3 s.f.)
   (d)   9.0909   (2 s.f.)
   (e)   0.00467 (1 s.f.)
   (f) 104.9   (2 s.f.)
   (g)   1.0449   (3 s.f.)

2 68 726 people attended a pop concert at Woburn.
Write this number correct to:
(a) 1 s.f.
(b) 3 s.f.
(c) 2 s.f.

3 Work out the following calculations. Give your answers correct
to the number of significant figures given in brackets:
(a) 439×49 (2 s.f.)
(b) $4.8 \div 0.313$ (4 s.f.)
(c) $976 \times 453 \times 182$ (1 s.f.)
(d) 0.00149×0.0837 (3 s.f.)
(e) $77 \times 88 \times 99$ (3 s.f.)

4 1 centimetre is approximately equal to 0.394 inches. Write:
(a) 1 foot (12 inches) in centimetres correct to 2 s.f.
(b) 1 yard (3 feet) in centimetres correct to 3 s.f.
(c) 1 mile (1760 yards) in centimetres correct to 1 s f

5 In each of parts **(a)** to **(d)**:
　(i) write down a calculation that could be used to estimate the
　　answer
　(ii) work out the estimated answer
　(iii) use a calculator to work out the exact answer

(a) 3.42×7.69
(b) $(1269 \times 176) + 402$
(c) $\dfrac{259.4 \times 46.9}{83.2}$
(d) $\dfrac{0.0456 \times 0.00296}{0.453}$

6 Estimate the answers to each of the following and give the
answers to the nearest whole number:
(a) $41 \div 8$　　　　　　　**(b)** $111 \div 12$
(c) $198 \div 43$　　　　　　**(d)** $287 \div 21$
(e) $497 \div 36$　　　　　　**(f)** 142×27
(g) $\dfrac{46 \times 33}{78}$　　　　　　**(h)** $\dfrac{54 \times 71}{36}$

7 Michael is working out the radius of a circle.
He has to do the calculation:

$$r = \sqrt{\dfrac{76.99}{3.142}}$$

Estimate, to the nearest whole number, r, the radius of the circle.

Exercise 12.3   Links: (*12E, F*) 12E, F

1 Copy and complete this table showing the attendance to the nearest hundred at these rugby stadiums.

Rugby stadium	Attendance to the nearest 100	Least possible attendance	Greatest possible attendance
Newcastle	1200		
Bedford	4600		
Saracens	11 700		
Leicester	13 200		
Bath	9800		

2 The distance between Hitchin and Shefford is given as 8 miles on a road sign.
Write down the range in which the true distance could lie.

3 These lengths are given to the nearest cm. Write down the smallest and largest possible lengths they could be:
 (a) 48 cm   **(b)** 26.43 m
 (c) 9.05 m   **(d)** 1.3842 km

4 The length of each side of a regular octagon is 5.6 cm correct to 2 significant figures. Calculate the greatest length the perimeter could be.

5 The speed of light is 186 000 miles per second to the nearest thousand miles per second.
The distance of the Earth from the Sun is 93.5 million miles correct to the nearest one hundred thousand miles.
A ray of light leaves the Sun and travels to the Earth.
It takes time *T*.
Calculate the range within which the time *T* taken by the ray lies.

6 *ABC* is a right-angled triangle.
 (a) Write down:
      (i) the upper bound of *BC*
      (ii) the lower bound of *BC*
 (b) Calculate:
      (i) the upper bound of the area of the triangle
      (ii) the lower bound of the area of the triangle

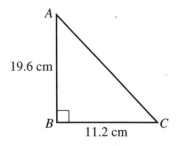

13 Basic trigonometry

Remember:
Check that your
calculator is working
in **degree mode**.

1 Use your calculator to find the value of:

   (a) $\sin 30°$ **(b)** $\cos 60°$ **(c)** $\tan 45°$ **(d)** $\sin 56°$

   (e) $\tan 78°$ **(f)** $\cos 54.6°$ **(g)** $\sin 38.1°$ **(h)** $\tan 32.5°$

2 Use your calculator to find the angle a when:

   (a) $\sin a = 0.866$ **(b)** $\tan a = 1.192$ **(c)** $\cos a = 0.5$

   (d) $\tan a = 0.603$ **(e)** $\sin a = 0.621$ **(f)** $\tan a = 0.6$

   (g) $\cos a = 0.75$ **(h)** $\sin a = 0.5$ **(i)** $\cos a = (\sqrt{2} \div 2)$

3 Calculate the size of the angles marked with a letter:

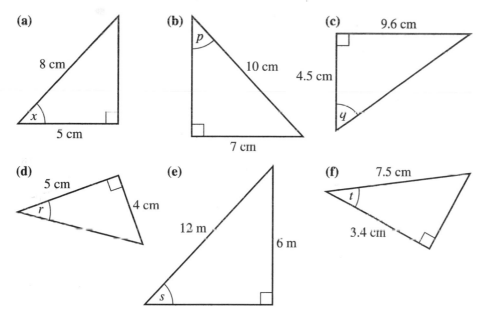

4 Calculate the length of the sides marked with a letter:

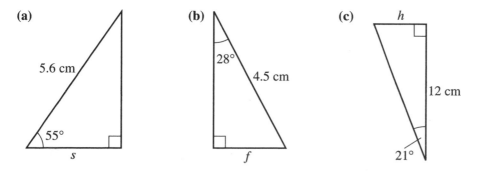

5 Calculate the size of the sides marked with a letter. All the lengths are in metres.

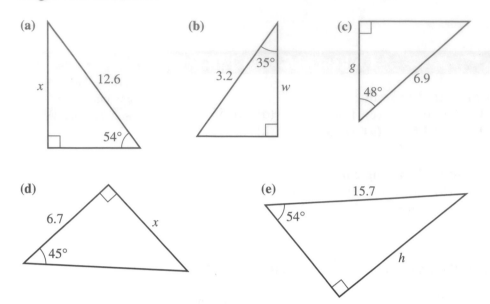

(a)

x 12.6   54°

(b)

3.2   35° w

(c)

g 48°   6.9

(d)

6.7   45° x

(e)

15.7   54° h

6 Bob places his 5-metres-long ladder, on horizontal ground, against the vertical wall of his house. The ladder makes an angle of 65° with the ground.
   (a) How far away from the wall is the foot of the ladder?
   (b) How far up the wall does the ladder reach?

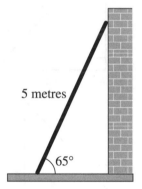

5 metres

65°

7 Work out the length of the sides marked with a letter in these right-angled triangles.

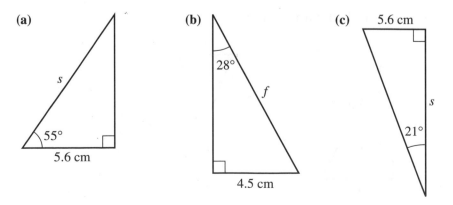

(a)

s 55°   5.6 cm

(b)

28° f 4.5 cm

(c)

5.6 cm s 21°

8 An isosceles triangle has two equal sides
   of length 10 centimetres. The other side
   has length 8 centimetres.
   Calculate the sizes of the angles
   of this triangle.

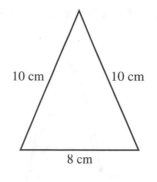

10 cm   10 cm

8 cm

9 An equilateral triangle has a vertical height of 15 cm. Calculate
   the length of the sides.

10 Calculate the size of the sides marked with a letter. All the
   lengths are in metres.

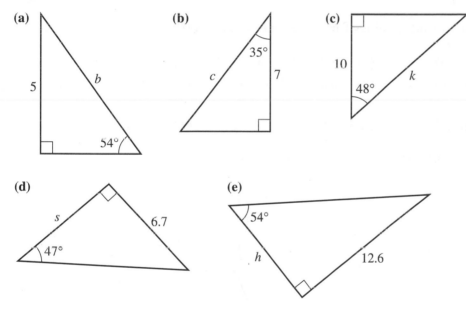

(a)

5

b

54°

(b)

35°

c   7

(c)

10

48°

k

(d)

s

6.7

47°

(e)

54°

h   12.6

Exercise 13.2   Links: *(13G – J)* 13G – J

1 Use your calculator to find the value of:
   (a) sin 120°   **(b)** cos 120°
   (c) tan 120°   **(d)** sin 210°
   (e) cos 210°   **(f)** tan 210°
   (g) cos 310°   **(h)** tan 310°
   (i) sin 310°   **(j)** sin −30°
   (k) cos −30°   **(l)** tan −30°
   (m) sin −300°   **(n)** cos −300°
   (o) tan −315°   **(p)** sin 180°

2 Write down the names of the 3 graphs in the diagram.

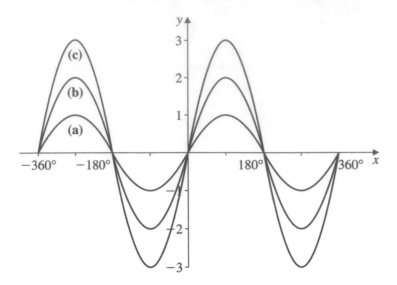

3 Use your calculator to find 4 values for angle b between $-360°$ and $360°$ in these cases:
   (a) $\sin b = 0.5$ **(b)** $\cos b = 0.5$ **(c)** $\tan b = 1$ **(d)** $\sin b = -0.5$
   (e) $\cos b = -\frac{1}{3}$ **(f)** $\tan b = -\frac{2}{3}$ **(g)** $\sin b = -\frac{1}{4}$ **(h)** $\cos b = 0$

4 On the same grid and for values of x between $-360°$ and $+360°$ draw the graphs of:
   (a) $\cos x$ **(b)** $\cos 2x$ **(c)** $\cos 3x$

5 This is the graph of
$$y = A \sin (x + B)$$
Use the graph to estimate the values of A and B.

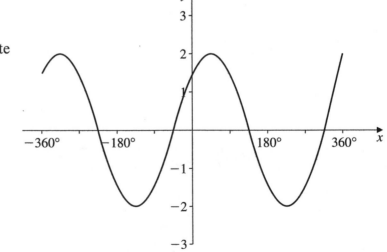

6 Solve the following trigonometric equations for values of x between $-360°$ and $+360°$.
   (a) $3 \sin 2x = 2$ **(b)** $2 \cos 3x = 1$ **(c)** $3 \tan 4x = 10$
   (d) $5 \sin 3x = 4$ **(e)** $10 \cos 2x = 3$ **(f)** $2 \tan 5x = 7$

7 Here is the graph of

$$y = 15\sin(x - 90°)$$

Make a copy of the graph and label the axes.

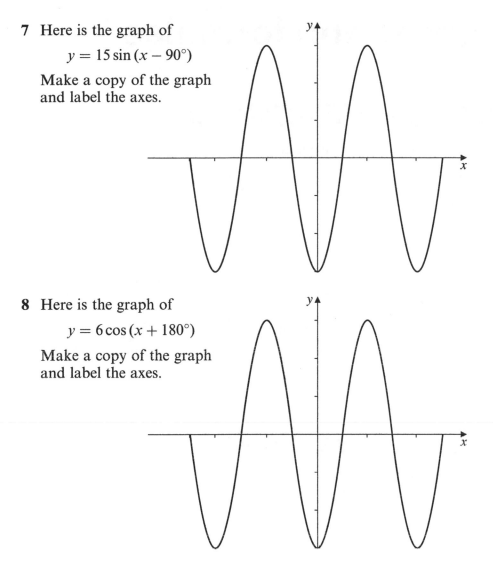

8 Here is the graph of

$$y = 6\cos(x + 180°)$$

Make a copy of the graph and label the axes.

9 The height h metres of water in a harbour at time t hours after midnight is given by the formula

$$h = 6 + 5\sin(30t)$$

   (a) Draw a sketch of the graph of h against t for values of t from 0 to 24.
   (b) What will be the
      (i) maximum depth of water in the harbour,
      (ii) minimum depth of water in the harbour.

There is an obstruction 2 metres high across the mouth of the harbour.

   (c) Between what times will a ship that needs a depth of 8 m of water be able to enter and leave the harbour.

10 Write down the maximum and minimum values of the function

$$12\cos(5x + 90°) - 5$$

14 Sequences and formulae

Exercise 14.1   Links: (*14A, B*) 14A, B

In questions **1–6** find the first five terms in the sequence.

1 $u_n = 3n$ **2** $u_n = n + 4$ **3** $u_n = 2n^2$

4 $u_n = 2n + 1$ **5** $u_n = n^2 - 1$ **6** $u_n = (2n + 1)^2$

In questions **7–15** find the value of n for which u_n has the stated value.

7 $u_n = 3n - 1$ $u_n = 59$ **8** $u_n = 6n - 3$ $u_n = 45$

9 $u_n = n^2 + 3$ $u_n = 124$ **10** $u_n = 2n^2$ $u_n = 162$

11 $u_n = n^3$ $u_n = 1728$ **12** $u_n = n(n + 1)$ $u_n = 42$

13 $u_n = \dfrac{2n + 1}{3n - 1}$ $u_n = \frac{7}{8}$ **14** $u_n = \dfrac{4 - n}{8 - 2n}$ $u_n = \frac{3}{2}$

15 $u_n = 2^{(n-4)}$ $u_n = 512$

Exercise 14.2   Links: (*14C*) 14C

Find the formulae for u_n to describe each of the following:

1 1, 4, 7, 10, 13 ...   **2** 3, 5, 7, 9, 11 ...

3 10, 15, 20, 25, 30 ...   **4** −3, −1, 1, 3 ...

5 17, 14, 11, 8, 5 ...   **6** 3, $3\frac{1}{2}$, 4, $4\frac{1}{2}$, 5 ...

7 0, 3, 8, 15, 24 ...   **8** 3, 9, 19, 33, 51 ...

9 4, 12, 36, 108, 324 ...   **10** 2, 10, 50, 250, 1250 ...

Exercise 14.3   Links: (*14D, E*) 14D, E

1 $A = LB$
   (a) Calculate the value of A when $L = 2.7$ and $B = 5.2$
   (b) Calculate B when $A = 32$ and $L = 4$

2 $P = 2(L + B)$
   (a) Calculate the value of P when $L = 1.07$ and $B = 4.23$
   (b) Calculate L when $P = 26.4$ and $B = 3.1$

3 $A = 2\pi r(r + h)$
   (a) Calculate the value of A when $r = 5.72$ and $h = 6.11$
   (b) Calculate h when $A = 122$ and $r = 4$

4 $x = L(1 + at)$
   (a) Calculate the value of x when $L = 120$, $a = 0.00002$ and $t = 50$
   (b) Calculate L when $x = 15.63$, $a = 0.000014$ and $t = 60$

5 $v = u + at$
   (a) Calculate the value of v when $u = 35$, $a = 6$ and $t = 5$
   (b) Calculate a when $v = 40$, $u = 24$, $t = 2$
   (c) Calculate t when $v = 14$, $u = 46$, $a = -5$

6 $I = \dfrac{PRT}{100}$
   (a) Calculate the value of I when $P = 2500$, $R = 6.25$ and $T = 2$
   (b) Calculate P when $I = 62.5$, $R = 5$, $T = 3$

7 $h = ut + \frac{1}{2}at^2$
   (a) Calculate the value of h when $u = 63.1$, $a = -9.81$ and $t = 3$
   (b) Calculate u when $h = 12$, $t = 1$ and $a = 4$
   (c) Calculate a when $h = 15.6$, $t = 3$, $u = 4.2$

8 $f = \dfrac{u + v}{uv}$
   (a) Calculate the value of f when $u = 6.4$ and $v = 3.6$
   (b) Calculate u when $f = 0.2$ and $v = 60$

9 $T = 2\pi\sqrt{\dfrac{l}{g}}$
   (a) Calculate correct to 3 s.f. the value of T when $l = 16.2$ and $g = 9.82$
   (b) Calculate l when $T = 0.45$ and $g = 9.8$

10 $T = \dfrac{2Mmg}{M + m}$
   (a) Calculate correct to 2 s.f. the value of T when $M = 5.6$, $m = 12$ and $g = 9.8$
   (b) Calculate g when $T = 23.4$, $M = 30$, $m = 20$
   (c) Calculate M when $T = 7.6$, $g = 9.8$, $m = 36$

Exercise 14.4   Links: (*14F*) 14F

Make the letter in brackets the subject of the formula:

1	$x = a(b + c)$	$[a]$	**2**	$x = a(b + c)$	$[b]$
3	$M = DV$	$[V]$	**4**	$E = mc^2$	$[c]$
5	$s = \frac{1}{2}gt^2$	$[g]$	**6**	$s = \frac{1}{2}gt^2$	$[t]$
7	$x = L(1 + at)$	$[L]$	**8**	$x = L(1 + at)$	$[a]$
9	$s = ut + \frac{1}{2}at^2$	$[a]$	**10**	$p = qr + q$	$[q]$

Exercise 14.5   Links: (*14G*) 14G

1 Mathew uses this formula to calculate the value of D:

$$D = \frac{a - 3c}{a - c^2}$$

Calculate the value of D when $a = 19.9$ and $c = 4.05$ [E]

2 The volume of a shape is given by the formula:

$$V = \tfrac{1}{3}\pi h(a^2 + ab + b^2)$$

Calculate the volume when $a = 15\,\text{cm}$, $b = 25\,\text{cm}$ and $h = 35.2\,\text{cm}$

3 $y = ab + c$

Calculate the value of y when $a = \tfrac{3}{4}$, $b = \tfrac{7}{8}$ and $c = -\tfrac{1}{2}$ [E]

4 The volume, V, of a barrel is given by the formula:

$$V = \pi H(2R^2 + r^2) \div 3000$$

$\pi = 3.14$, $H = 60$, $R = 25$ and $r = 20$.
Calculate the value of V correct to 3 significant figures.   [E]

5 For the formula $W = \dfrac{\lambda x^2}{2L}$

Make **(a)** L the subject and   **(b)** x the subject

6 The formula used for working out the range, R metres, of a gun firing shells with a muzzle velocity, $v\,\text{ms}^{-1}$, from a cliff, height h metres is

$$R = \frac{v}{g}\sqrt{v^2 + 2gh}$$

The formula is used when $v = 215\,\text{ms}^{-1}$, $h = 53\,\text{m}$ and $g = 9.8\,\text{ms}^{-2}$
(a) Without using a calculator and showing all your working quote sensible approximations for v, h and g to calculate an estimate for R.
(b) Calculate the actual range.
(c) Calculate the range when the gun is made 10% more powerful.
(d) What is the percentage increase in range?

15 Averages and measures of spread

1 The Art coursework marks for twelve students marked out of 80 were:

 51, 54, 57, 38, 63, 49, 54, 72, 76, 28, 46, 54

 (a) Work out the:
   (i) mean   (ii) median   (iii) mode
   for these marks.
 (b) Comment on your results.
 (c) The moderator raises all the marks by 4. Work out the mean, median and mode of these marks.

2 The diagram represents a wheel which is spun about the centre.
 The number the arrow points to is the number recorded.
 The spinner is spun several times.
 (a) What is the theoretical mean of the recorded numbers?
 (b) Explain your result.
 (c) 'The mean is not a sensible measure of average'. Explain why.
 (d) What are the mode and median of the numbers? Explain your results.

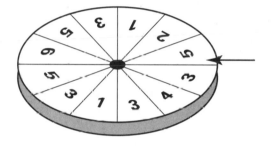

3 Sebastian wishes to conduct a survey on the amount of pocket money teenagers receive. Explain, with reasons, which averages he should use and why.

4 Find the mean, median and mode of these sets of data.
 (a) 16, 20, 28, 18, 24
 (b) 1007, 1001, 1005, 1016, 1005
 (c) $3y, 2y, 4y, 2y, y, 8y$

5 The mean of 6 numbers is 12.
 5 of the numbers are 10, 8, 14, 6, 14.
 Find the sixth number.

6 Alicia has information about the quarterly gas bill at her home over a four year period from 1998 to 2001. The information is given in the table below:

YEAR	February	May	August	November
1998	252	203	112	160
1999	264	210	118	163
2000	273	215	124	172
2001	294	224	131	189

   (a) Represent this data as a time series.
   (b) Calculate the four-point moving average and plot these on the same axes as the original data.
   (c) Complete the trend line for the moving averages.
   (d) Comment as fully as possible on the changes to the gas bill over the four-year period.

7 The average house price (£1000s) in Shimpwell from 1985 to 2001 is given below.

Year	85	86	87	88	89	90	91	92	93	94	95	96	97	98	99	00	01
Price	47	47	56	63	48	48	49	48	52	54	57	62	68	75	90	98	105

   (a) Plot the data as a time series.
   (b) Work out the 5-point moving average, plotting these at the trend line on the same axes as the time series.
   (c) Comment on the data and your graphs.

8 Ten adults did a crossword. The mean time was 13.4 minutes. A different twenty five adults did the same crossword and had a mean time of 16.5 minutes.
   (a) Calculate the mean time for all 35 adults.
   (b) The median for the ten adults was 13.8 and the median for the twenty five adults was 17.1.
      What can you say about the median for all 35 adults?

9 The marks of 100 students are shown in the table.
   (a) Calculate the mean.
   (b) Calculate the median.
   (c) Write down the mode.

Mark	Frequency
1	0
2	1
3	8
4	9
5	35
6	23
7	16
8	5
9	2
10	1

10 Melanie did a survey to record the number of pets her classmates had in their household
  (a) Calculate the mean.
  (b) Calculate the median.
  (c) State the mode.

Number of pets	Number of households
0	3
1	7
2	4
3	6
4	5
5	2
6	2
7	1

Exercise 15.2   Links: (*15C, D*) 15C, D, E

1 The lifetime in hours of 100 batteries are shown in the table.

Time t (hours)	Frequency
$0 \leqslant t < 3$	2
$3 \leqslant t < 6$	3
$6 \leqslant t < 9$	16
$9 \leqslant t < 12$	34
$12 \leqslant t < 15$	27
$15 \leqslant t < 18$	13
$18 \leqslant t < 21$	5

  (a) Calculate an estimate of the mean.
  (b) Write down the class interval in which the median lies.
  (c) State the modal class.

2 The lengths of their journeys to school were recorded for 200 pupils at Harkspar School.

Length of journey (km)	Frequency
0 and less than 1	11
1 and less than 2	34
2 and less than 3	51
3 and less than 4	63
4 and less than 5	19
5 and less than 6	15
6 and less than 7	7

  (a) Calculate an estimate of the mean.
  (b) State the modal class.
  (c) Write down the class interval in which the median lies.

3 For the data in questions **1** and **2** calculate estimates of the following values by drawing a cumulative frequency graph.
  (a) median   **(b)** lower quartile
  (c) upper quartile   **(d)** interquartile range
  (e) Draw the box and whisker diagram.

4   The weekly wage of 100 workers in East Anglia is given below.

Wage (£)	Frequency
12–140	2
141–160	6
161–180	12
181–200	38
201–220	29
221–240	9
241–260	4

By drawing a cumulative frequency graph calculate:
(a) an estimate of the median
(b) an estimate of the interquartile range
(c) the percentage of workers earning less than £172.
(d) Draw the box and whisker diagram.

5   The weights of 80 pigs were measured at a farm.

Weight (kg)	Number of pigs
40 to under 45	3
45 to under 50	6
50 to under 55	17
55 to under 60	27
60 to under 65	14
65 to under 70	9
70 to under 75	4

(a) Calculate an estimate of the mean.

By drawing a cumulative frequency graph, estimate:
(b) the median   (c) the upper quartile
(d) the interquartile range   (e) the percentage of pigs less than 62 kg.
(d) Draw the box and whisker diagram.

6   A safari park game keeper records the ages of 200 elephants in
his park. The results are shown in the table:

Age (years')	0–19	20–39	40–59	60–79	80–99	100–119	120–139	140–159
Frequency	7	11	18	37	46	36	31	14

(a) Calculate an estimate of the mean age.

By drawing a cumulative frequency graph, estimate:
(b) the median
(c) the upper quartile
(d) the interquartile range
(e) how many elephants are at least 110 years old
(f) the percentage of elephants less than 65 years old.

7 The diagrams represent the histograms for three distributions: In each case, sketch the cumulative frequency curve.

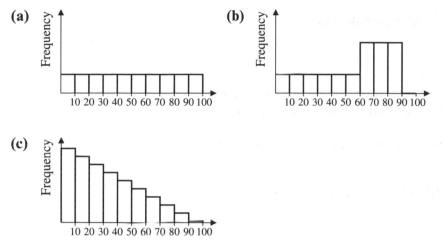

(a)
Frequency
10 20 30 40 50 60 70 80 90 100

(b)
Frequency
10 20 30 40 50 60 70 80 90 100

(c)
Frequency
10 20 30 40 50 60 70 80 90 100

8 The Local Education Authority wants to compare the LEA test results in two schools. They gathered the following information:

	School A	School B
Maximum %	15	30
Minimum %	98	96
Median	58	72
Upper quartile %	76	84
Lower quartile %	28	40

(a) Choose an appropriate scale and draw the box and whisker diagram for each school.
(b) Which school do you believe had the better results? Give your reasons.

Exercise 15.3   Links: (*15E*) 15G

1 The year 11 pupils at Harkspar High School sat an exam in geography. The exam consisted of two papers A and B. The table shows the medians and interquartile ranges of the marks of the two papers.

Paper	Median (%)	Interquartile range
A	61	16
B	82	23

Comment on this data.

2 The table shows the medians and interquartile ranges of the
   weights of two rugby teams.

Team	Median (kg)	Interquartile range
A	89	17
B	95	14

Explain whether or not it would be fair to say that members of
club B are in general heavier than members of club A.

3 In an examination consisting of two papers, the following
   statistics were collected:

Paper	Median % mark	Interquartile range
1	54	28
2	71	15

Make statistical comparisons between Paper 1 and Paper 2.

16 Measure and mensuration

1 A table mat measures 25 cm by 20 cm.
Its dimensions are quoted to the
nearest centimetre.
   (a) Write down the shortest and
   longest widths of the mat.
   (b) Write down the shortest and
   longest lengths of the mat.
   (c) Calculate the smallest and
   largest areas of the mat.

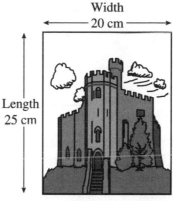

Width
20 cm

Length
25 cm

2

(2x + 1) cm

(x + 3) cm

   (a) Find in its simplest form an expression in x for the perimeter
   of this rectangle.
   (b) Given the perimeter is 104 cm, calculate the value of x.
   (c) Calculate the area of the rectangle.

3 The length of a rectangle is twice the width. The area of the
rectangle is 162 cm². Calculate the perimeter of the rectangle.

1 Calculate the areas of these triangles:

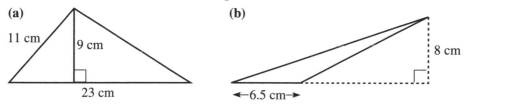

   (a)

11 cm 9 cm

23 cm

   (b)

8 cm

←6.5 cm→

2 The perimeter of this triangle is 32 cm.
   (a) Calculate the value of x.
   (b) Write down the lengths of the 3 sides.

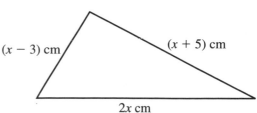

(x − 3) cm (x + 5) cm

2x cm

3 Calculate the areas of these parallelograms:

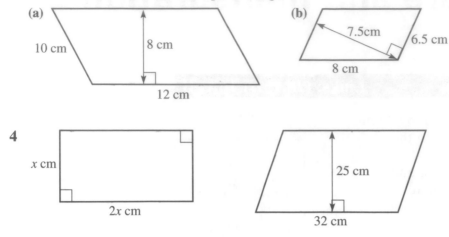

(a)

10 cm

8 cm

12 cm

(b)

7.5cm

6.5 cm

8 cm

4

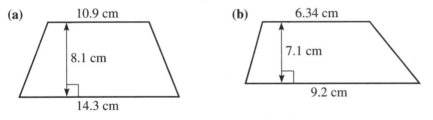

x cm

2*x* cm

25 cm

32 cm

The diagram shows a rectangle and a parallelogram. These two shapes have equal areas.

(a) Calculate the value of *x*.

(b) Write down the lengths of the two sides of the rectangle.

5 Calculate the areas of these trapeziums:

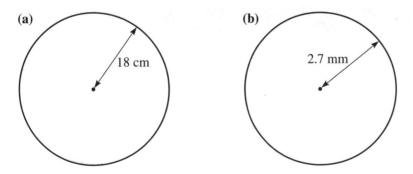

(a)   10.9 cm

8.1 cm

14.3 cm

(b)   6.34 cm

7.1 cm

9.2 cm

Exercise 16.3   Links: (*16F*) 16F

1 Calculate the circumferences and areas of these circles:

(a)

18 cm

(b)

2.7 mm

2 The area of a circle is 150 cm².
   Calculate the circumference of the circle.

3 The circumference of a circle is 68 cm.
   Calculate the area of the circle.

4 The diagram represents a running track.
It consists of a rectangle and two semi-circles.

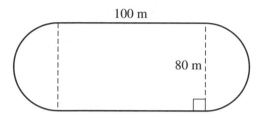

Calculate:
(a) the perimeter of the running track
(b) the area enclosed by the running track

Exercise 16.4   Links: (*16G, H*) 16G, H

1 Calculate the volume of a cube of side length:
(a) 6 cm
(b) 4.3 cm
(c) 29.8 mm

2 A cube has a volume of 729 cm^3.
Calculate:
(a) the length of the side of the cube
(b) the surface area of the cube

3 Calculate the volumes and surface areas of the following cuboids:

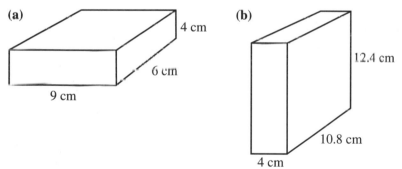

4 Calculate the volumes and surface areas of these prisms:

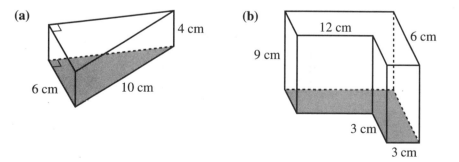

5 The prism *ABCDEF* has volume $= 510\,\text{cm}^3$,
   $BC = 10\,\text{cm}$, $DC = 12\,\text{cm}$.
   Calculate the length *AB*.

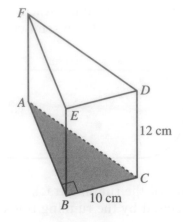

Exercise 16.5   **Links:** *(16I, J)* 16I, J

1 Here are two expressions related to a cone of radius *r*, height *h*
   and slant height *l*:

   $\pi r l$ and $\frac{1}{3}\pi r^2 h$

   State, with reasons, which expression gives the volume of the
   cone and which gives the curved surface area.

2 In the following expressions λ is dimensionless and *a*, *b*, *c* have
   dimensions of length.
   Explain whether each expression represents a length, an area or
   a volume:

   $a\lambda^2 b$, $abc\lambda$, λa, $\lambda a^2 c$, $\lambda(a + b + c)$

3 Calculate the surface area and volume of the wedge *ABCDEF*
   in which $AB = 6\,\text{cm}$, $BC = 10\,\text{cm}$ and $CD = 15\,\text{cm}$.

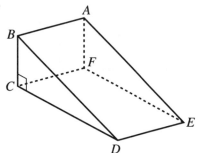

4 Calculate the surface area and volume
   of this trapezoidal prism.

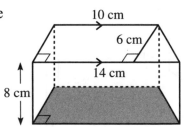

17 Proportion

1 Which of the following could be examples of direct proportion:
   (a) the number of pens bought and the cost
   (b) the cost of a used car and the age of the car
   (c) the volume of a cube and the length of one edge
   (d) the speed of a car and the distance travelled in one hour
   (e) the time spent on your maths course-work and the final
       grade awarded for it
   (f) the engine size of a car and its fuel consumption.

2 When $y \propto x$ and $y = 5$ when $x = 12$, find the value of:
   (a) x when the value of y is doubled
   (b) y when the value of x is divided by 3

3 When $y \propto x$ and $y = 12$ when $x = 9$, find the value of:
   (a) x when $y = 72$ **(b)** y when $x = 3$

4 The length of a shadow s cast by a tree is
   directly proportional to the height h of the
   tree. At 2 pm a tree of height 30 metres casts
   a shadow of 18 metres.
   (a) Sketch the graph of the relationship.
   (b) Find the height h of another tree at 2 pm
       when the length of the shadow is 15 metres.
   (c) Find the length s of the shadow
       at 2 pm of a tree that has a height
       of 40 metres.

5 The amount of electricity w that is used is directly proportional
   to the current used a. When $w = 600$, $a = 2.5$.
   (a) Sketch a graph to show this information.
   (b) Write down a rule that connects w and a.
   (c) When $a = 4$ find the value of w.
   (d) When $w = 1000$ find the value of a.

6 When $y \propto x$ and $y = 1.2$ when $x = 0.5$, find the value, to 3
   significant figures, of:
   (a) x when $y = 8.9$ **(b)** y when $x = 0.3$

7 p varies directly as q so that $p \propto q$. When $p = 6$, $q = 8$.
   (a) Find the value of p when $q = 24$.
   (b) Find the value of q when $p = 42$.

8 t varies directly as r. When $t = 8.4$, $r = 3.6$.
Find the value of:
(a) t when $r = 4.8$ (b) r when $t = 9.8$

9 The circumference C of a circle varies directly as the radius r.
When the circumference is 22 cm the radius is 3.5 cm. Find the value of:
(a) C when $r = 5.6$ cm   (b) r when $C = 87$ cm

10 The volume V of water in a cylinder varies directly as the height h. When the height of the water in the cylinder is 12 cm the volume is 50 ml.
(a) Find the volume of water when the height is 15 cm.
(b) Find the height of water when the volume is 80 ml.

Exercise 17.2   Links: (*17E*) 17E

1 y is directly proportional to the square of x, so that $y = kx^2$.
Given that $y = 36$ when $x = 3$:
(a) calculate the value of k
(b) calculate the value of y when $x = 5$
(c) calculate the value of x when $y = 64$

2 p is directly proportional to the cube of q, so that $p = kq^3$.
Given that $p = 24$ when $q = 2$:
(a) calculate the value of k
(b) calculate the value of p when $q = 3$
(c) calculate the value of q when $p = 375$

3 A varies in direct proportion to the square of r. Given that when $r = 4$, $A = 50$:
(a) calculate the value of A when $r = 3$
(b) calculate the value of r when $A = 100$

4 V varies in direct proportion to the cube of r. Given that when $r = 3$, $V = 113$:
(a) calculate the value of V when $r = 5$
(b) calculate the value of r when $V = 40$

5 The speed S of a car is directly proportional to the square of the braking power P. When the braking power is 35 the speed is 20.
(a) Calculate the value of S when $P = 50$.
(b) Calculate the value of P when $S = 70$.

6 The volume V of a hemisphere varies directly as the cube of the radius r. When the radius is 5 cm the volume is 262 cm^3.
(a) Calculate the volume when the radius is 2 cm.
(b) Calculate the radius when the volume is 50 cm^3.

7 A stone is dropped from the top of a cliff. The distance of the stone from the top varies as the square of the speed. When the speed is 10 metres per second the distance is 5 metres.
   (a) Calculate the distance from the top when the speed is 12 metres per second.
   (b) Calculate the speed when the distance from the top is 30 m.

8 The surface area of a cylinder is directly proportional to the square of the radius. When the area is $12\,cm^2$ the radius is 2.5 cm. Calculate the area when the radius is 5 cm.

Exercise 17.3   Links: (*17F*) 17F

1 y is inversely proportional to x so that $y = \dfrac{k}{x}$

When $x = 10$ the value of y is 2.
   (a) Find the value of k.
   (b) Find the value of y when $x = 40$.
   (c) Find the value of x when $y = 10$.

2 p is inversely proportional to q so that $p = \dfrac{k}{q}$

When $p = 2$ the value of q is 0.01.
   (a) Find the value of p when $q = 4$.
   (b) Find the value of q when $p = 0.1$.

3 r is inversely proportional to the square of t so that $r = \dfrac{k}{t^2}$

When $r = 8$ the value of t is 0.5.
   (a) Find the value of r when $t = 3$.
   (b) Find the value of t when $r = 10$.

4 The gravitational pull F of a planet is inversely proportional to the square of the distance d from the planet. When the distance from the planet is 1000 km the gravitational pull is 12 N.
   (a) Calculate the gravitational pull when the distance is 200 km.
   (b) Calculate the distance when the gravitational pull is 50 N.

5 The luminescence L of a light source is inversely proportional to the square of the distance d from the light source. When the distance from the light source is 2 metres the luminescence is 5 lumens.
   (a) Calculate the luminescence when the distance is 5 metres.
   (b) Calculate the distance when the luminescence is 20 lumens.

6 y is inversely proportional to the square root of x.
   When $y = 5$, $x = 12$.
   (a) Calculate the value of y when $x = 20$.
   (b) Calculate the value of x when $y = 3$.

7 The dispersion factor D of molecules of gas in a spherical
container is inversely proportional to the cube of the radius r of
the sphere. When the radius is $10\,cm$ the dispersion factor is 4.
   (a) Calculate the dispersion factor D when the radius is $5\,cm$.
   (b) Calculate the radius r when the dispersion factor is 20.

8 The loudness L of a sound is inversely proportional to the
square root of the distance d from the source of the sound.
When the distance is 20 metres the loudness is 90 decibels.
   (a) Calculate the loudness L when the distance is 10 metres.
   (b) Calculate the distance when the loudness is 120 decibels.

Exercise 17.4   Links: (*17G*) 17G

1 If $y \propto x$ and $y = 15$ when $x = 3$, find the value of
   (i) y when $x = 2$,
   (ii) x when $y = 20$.

2 If $y \propto x$ and $y = 8$ when $x = 10$, find the value of
   (i) y when $x = 100$,
   (ii) x when $y = 100$.

3 The table gives values of x and y.
   Which of these expressions could be true?

x	1	5	10	20
y	5	125	500	2000

$y \propto x$, $y \propto x^2$ $y \propto \dfrac{1}{x}$,

$y = 5x$, $y = 5x^2$, $y = 2x + 3$, $y = x^3$

4 If $y \propto x^3$ and $y = 24$ when $x = 2$, find the value of
   (i) y when $x = 7$,
   (ii) x when $y = 192$.

5 If y is inversely proportional to the square of x and $y = 8$ when
$x = 0.5$, find the value of
   (i) y when $x = 10$,
   (ii) x when $y = 200$.

6 Work out the relationships between y and x in these tables

   (a)

x	1	2	3	4
y	15	30	45	60

   (b)

x	1	2	3	4
y	10	5	$3\frac{1}{3}$	$2\frac{1}{2}$

   (c)

x	1	2	3	4
y	8	32	72	128

18 Graphs and higher order equations

(a) Draw graphs with the following equations, taking values of x from -1 to 7.

(b) In each case give the coordinates of the vertex.

1 $y = x^2 - 5x$ 2 $y = x^2 - 6x + 1$

3 $y = (x - 3)^2 - 4$ 4 $y = 2x^2 - 3x + 15$

5 $y = 3x^2 - 16x + 16$ 6 $y = (x - 2)^2 + 18$

7 $y = 2x^2 - 3x - 19$ 8 $y = (2 - x)^2 - 3$

9 $y = (3 - x)^2 - 6$ 10 $y = (4 - x)^2 + 5$

In questions **11–20** take values of x from -1 to 5.

11 $y = x^3 - 4x^2 - 8$ 12 $y = 2x^3 - 12x^2 + 12x - 16$

13 $y = x^3 - 6x^2 + 6x - 18$ 14 $y = x^3 - 3x^2 + 3x - 1$

15 $y = x^3 - 6x^2$ 16 $y = x^3 - 6x^2 + x + 2$

17 $y = x^3 - 6x^2 + 9x - 16$ 18 $y = x^3 - 5x^2 - 6x + 12$

19 $y = -x^3 + 7x^2 - 10x + 12$ 20 $y = 8 - 6x + 6x^2 - x^3$

In questions **21–26** take values of x from -3 to 3.

21 $y = x^3 - 12x$ 22 $y = x^3 + 3x^2 + 2$

23 $y = x^3 - 3x^2 + 4$ 24 $y = x^3 - 6x + 2$

25 $y = 2 + 12x - x^3$ 26 $y = 3x^2 - x^3$

Draw graphs with the following equations using values of x from -1 to 5.

1 $y = \dfrac{10}{x}$ 2 $y = 1 - \dfrac{5}{x}$ 3 $y = \dfrac{1}{x - 2}$

4 $y = \dfrac{1}{1 - x}$ 5 $y = 2 + \dfrac{1}{x - 1}$

6 $y = \dfrac{x - 1}{x - 2}$ 7 $y = \dfrac{2}{x - 3}$

In the following questions use the values of x shown in the brackets.

8 $y = x^3 + 3x^2 + \dfrac{1}{x}$ $(-4 \text{ to } 2)$

9 $y = 2x^3 - 5x - 1$ $(-2 \text{ to } 2)$

10 $y = 4x - \dfrac{1}{x}$ $(-3 \text{ to } 3)$

11 $y = x^2 + \dfrac{1}{x}$ $(-3 \text{ to } 3)$

12 $y = (x - 1)^3 - \dfrac{1}{1 - x}$ $(-2 \text{ to } 4, \text{ include } 0.9 \text{ and } 1.1)$

13 $y = x^2 + 2x + \dfrac{1}{x}$ $(-3 \text{ to } 3)$

14 $y = x^3 + x^2 - x$ $(-3 \text{ to } 3, \text{ include } 0.5)$

Exercise 18.3   Links: (*18E*) 18E

Solve the following equations correct to 1 d.p. by drawing appropriate graphs. Use the values of x given in the brackets.

1 $x^2 - 5x + 3 = 0$ $(0 \text{ to } 5)$

2 $2x^2 - 9x + 5 = 0$ $(0 \text{ to } 5)$

3 $3x^2 + 2x - 6 = 0$ $(-4 \text{ to } 3)$

4 $x^2 + 8x + 14 = 0$ $(-7 \text{ to } -1)$

5 $x^3 - x^2 - 3x = 0$ $(-3 \text{ to } 3)$ [3 solutions]

6 $x^2 + \dfrac{1}{x} - 4 = 0$ $(-3 \text{ to } 3)$ [3 solutions]

7 $x^2 + 2x - \dfrac{1}{x} = 0$ $(-3 \text{ to } 2)$ [3 solutions]

8 $x - \dfrac{1}{x} + 1 = 0$ $(-3 \text{ to } 2)$ [2 solutions]

Exercise 18.4   Links: (*18F*) 18F

Find the solution of these equations that is between the stated limits using trial and improvement. Give your answers to 2 d.p.

1 $x^3 + x = 37$ $(x = 3 \text{ and } x = 4)$

2 $x^3 - x = 20$ $(x = 2 \text{ and } x = 3)$

3 $x^3 + 3x = 15$ $(x = 2 \text{ and } x = 3)$

4 $(x - 7)^3 + x - 13 = 0$ $(x = 8 \text{ and } x = 9)$

5 $2x^3 + 3x = 10$ $\qquad$ ($x = 1$ and $x = 2$)

6 $x^2 = \dfrac{1}{x} + 1$ $\qquad$ ($x = 1$ and $x = 2$)

7 $x^2 + \dfrac{3}{x} = 7$ $\qquad$ ($x = 2$ and $x = 3$)

8 $x^2(2x + 1) - 27 = 0$ $\quad$ ($x = 2$ and $x = 3$)

9 $x(x - 1)(x - 2) = 15$ $\quad$ ($x = 3$ and $x = 4$)

10 $(2 - x)x^2 + 40 = 0$ $\quad$ ($x = 4$ and $x = 5$)

Exercise 18.5 $\qquad$ Links: (*18G*) 18G

1 The graph shows a short car journey.
  (a) Describe what is happening on the journey between:
    (i) A and B
    (ii) B and C
    (iii) C and D
  (b) What is the distance travelled in the first 5 seconds?
  (c) How long does it take to travel the first 40 metres?

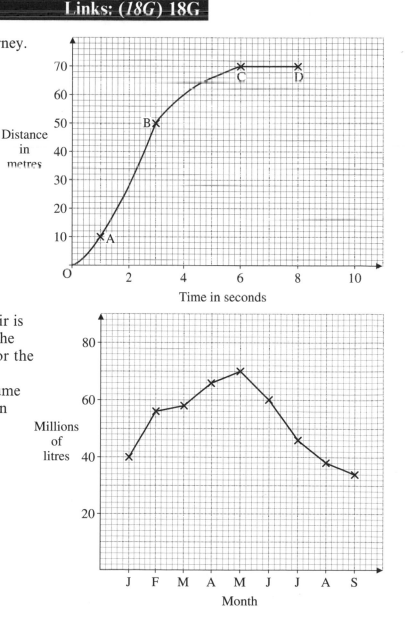

2 The volume of water in a reservoir is measured monthly over a year. The graph shows the measurements for the first nine months.
  (a) Describe briefly how the volume of the reservoir varies between January and August.
  (b) Give possible reasons for the graphical cycle.

3 The graph shows the level in a gas holder at different times
   during a certain day.

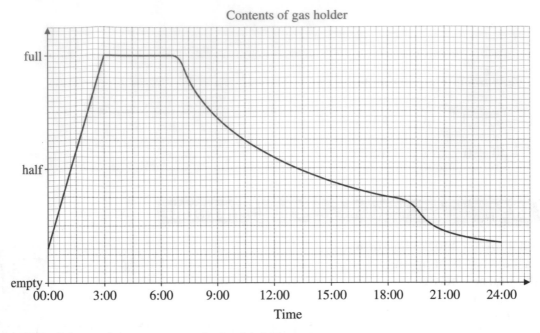

   (a) Estimate when gas usage is the highest.
   (b) Describe what is happening between 00:00 and 03:00.

4 The graph shows the number of people in a football ground.

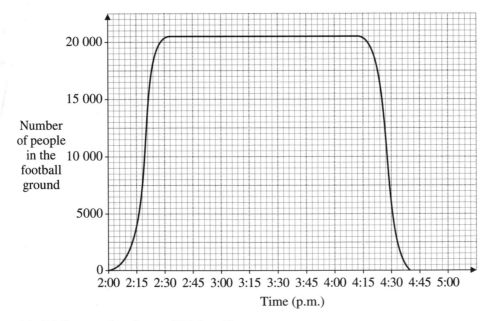

   (a) Estimate the time of kick-off.
   (b) Estimate the size of the football crowd.
   (c) Describe what is happening between 4:15 and 4:45.

5 The graph shows the temperature in an oven:

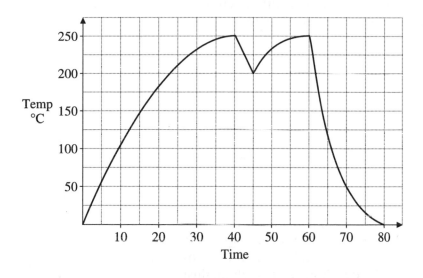

(a) What temperature was the oven set at?
(b) Give a possible explanation for what happened at between 40 and 50 minutes.
(c) When did cooking finish?

6 The graph shows the amount of water in a water butt which collects water from the roof of the shed.

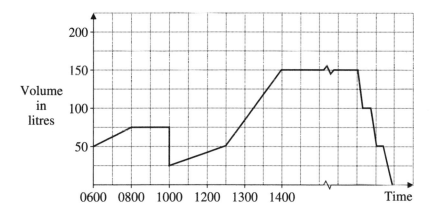

Between 6.00 am and 8.00 am there is light rain.
(a) Describe what happens between 10.00 am and 1.30 pm.
(b) It continues to rain until 3.00 pm.
What can you deduce?

7   Sand flows out of an egg timer in the shape of a cylinder on top
    of an inverted cone. Sketch a graph to show depth of sand
    against time as the sand flows out at a constant rate.

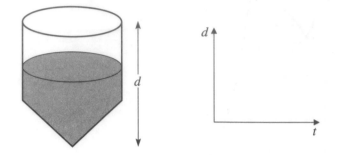

8   A horse trough is in the shape of a triangular prism. It is filled
    by a hose delivering 5 litres per second. After 12 seconds another
    hose, also delivering 5 litres per second, is added until the trough
    is full. Sketch a graph to show the depth in the trough against
    time in seconds.

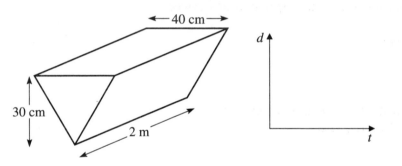

19 Advanced mensuration

In this exercise, give your answers to 3 significant figures where appropriate.

1 Calculate the arc length and the area of the sectors of these circles.

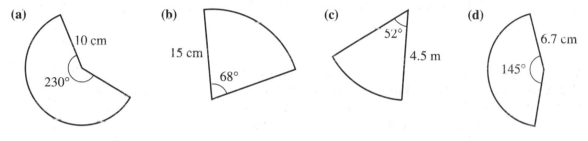

(a) 10 cm, 230°   **(b)** 15 cm, 68°   **(c)** 52°, 4.5 m   **(d)** 6.7 cm, 145°

2 Calculate the size of the angles marked *a* in these sectors of circles.

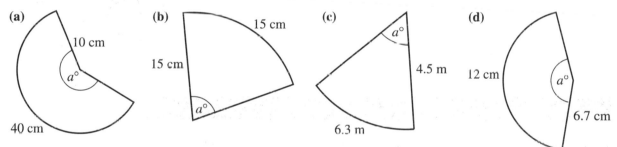

(a) 10 cm, $a°$, 40 cm   **(b)** 15 cm, 15 cm, $a°$ **(c)** $a°$, 4.5 m, 6.3 m   **(d)** 12 cm, $a°$, 6.7 cm

3 The area of a sector of a circle of radius 5 cm is 10 cm². Calculate the angle at the centre of the sector.

4 The arc length of the sector of a circle is 30 cm when the angle at the centre is 150°. Calculate the radius of the circle.

5 Calculate the shaded areas in these sectors.

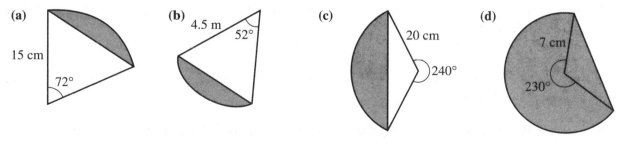

(a) 15 cm, 72°   **(b)** 4.5 m, 52°   **(c)** 20 cm, 240°   **(d)** 7 cm, 230°

6 A shot-put competition takes place in a circle of
   diameter 2 metres and a landing area that is the sector of
   a circle with radius 20 metres and an angle of 38°. The
   centre of both circles is the same. Calculate the total
   area of the sports field taken up by this shape.

7 The entrance to a tunnel on a canal is in the shape of a rectangle
   and the sector of a circle. The width of the rectangle is
   3.36 metres and the height is 3.22 metres. The centre of the arc
   of the circle is on the water line of the canal and has an angle of
   80°. The depth of the water in the canal is 1.22 metres. Calculate
   the perimeter of the tunnel entrance that is above the water line.

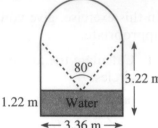

8 A stained glass window in a church is made up of sectors
   of circles. The outer circle has a radius of 2 metres. Each
   arc has a radius of 2 m, centre on the outer circle, and an
   angle of 90° at its centre. Each central arc is a quarter
   circle. The circumference and all the inner arcs represent
   the lead beading that holds the coloured glass in place.
   Calculate the length of lead beading used in this window.

Exercise 19.2   Links: (*19E, F, G*) 19E, F, G

In this exercise, give your answers to 2 decimal places where appropriate.

1 A cylindrical tin of fruit has a base radius of 3 cm and a height of 12 cm.
   (a) Calculate the volume of the tin of fruit.
   (b) Calculate the surface area of the tin.

2 A chocolate bar of length 15 cm is in the shape of a triangular
   prism. The triangular ends are equilateral triangles with edges of 5 cm.
   (a) Calculate the volume of the chocolate bar.
   (b) Calculate the surface area of the bar.

3 A hexagonal steel bar has a length of 30 cm. The length of each edge of the hexagonal end
   is 0.5 cm. Calculate:
   (a) the volume of the bar   (b) the surface area of the bar

4 A cylindrical waste bin has a volume of 40 litres. The radius of the base is 15 cm. Calculate
   the height of the bin.

5 Calculate the volume of a cone that has a base radius of 10 cm and a height of 15 cm.

6 A square-based pyramid has a height of 20 cm. The length of each edge of the square base is 10 cm. Calculate the volume of the pyramid in litres.

7 A solid cube of steel has a volume of 1 litre. The cube is melted down and recast as a cylinder of radius 5 cm. Work out the length of the cylinder.

8 A cone has a volume of 500 cm³ and a height of 12 cm. Calculate the radius of the base.

9 Calculate the volume and surface area of a sphere of:
   (a) radius 5 cm   (b) radius 0.5 cm   (c) radius 1 m

10 A hexagonal steel bar has a volume of 12 cm³. The length of the bar is 10 cm. Calculate the length of one edge of the hexagonal end.

11 Calculate the volume of a cube that has a surface area of 12 cm².

12 Calculate the volume of a sphere that has a surface area of 1000 cm².

Exercise 19.3   Links: (*19H*) 19H

In this exercise, give your answers to 3 s.f. where appropriate.

1 Terry makes a model of a World War II Spitfire. He uses a scale of 1 : 100. The wingspan of the real plane is 16 metres.
   (a) Calculate the wingspan of the model plane.

   The area of one of the circular markings on the real plane is 600 cm².
   (b) Calculate the area of one of the circular markings on the model plane.

   The volume of a fuel tank in the real plane was 5000 litres.
   (c) Calculate the volume of a fuel tank in the model plane.

2 A manufacturer of wheelbarrows makes them in 3 sizes; small, medium and large. The wheelbarrows are similar in shape and size. The ratio of the lengths of the wheelbarrows is 2 : 3 : 4.
   (a) If the area of the top of the medium wheelbarrow is 1 m² calculate the areas of the tops of the other wheelbarrows.
   (b) The small wheelbarrow will hold 0.6 m³. How much do the other two wheelbarrows hold?

   The radius of the wheel of the medium wheelbarrow is 10 cm.
   The volume of the wheel of the large wheel barrow is 3 litres.
   (c) Calculate the area of cross-section of the wheels of all 3 wheelbarrows.
   (d) Calculate the volumes of the wheels of the medium and small wheelbarrows.

3 The manufacturers of bottles of a soft drink make it in 1-litre, 2-litre and 3-litre bottles. All the bottles are similar in shape.
   (a) Calculate the ratio of the heights of the 3 bottles.
   (b) Calculate the ratio of the surface areas of the 3 bottles.

4 Two similar bottles of 'Sunshine Cola' have lengths that are 12 cm and 15 cm.
   (a) The volume of the large bottle is 1.5 litres. Calculate the volume of the smaller bottle.
   (b) The area of the label on the small bottle is 12 cm^2. Calculate the area on the label of the large bottle.

5 Mr Green sells ice cream cones that have similar shapes. The diameter of the small cone is 5 cm and the diameter of the large cone is 8 cm. Each cone is filled with ice cream.
   (a) If the volume of ice cream in the small cone is 100 cm^3 calculate the volume of ice cream in the large cone.
   (b) If the surface area of the large cone is 268 cm^2 what is the surface area of the small cone?

6 A scale model of a ship is built on a scale of 1:200. Copy this table into your book and complete the information.

Measurement	Ship	Model of ship
Height of mast	30 m	......cm
Area of deck		9.5 cm^2
Volume of hold	15 000 m^3	
Number of portholes	2000	

7 The height of the whole cone is 15 cm. An identical cone has a small cone at the top removed leaving the bottom section (frustum). The height of this bottom section is 10 cm. The surface area of the small cone is 30 cm^2. The volume of the large cone is 250 cm^3.
   (a) What is the scale factor of the lengths of the large cone to the small cone?
   (b) Work out the surface area of the large cone.
   (c) Calculate the volume of the frustum of the cone.

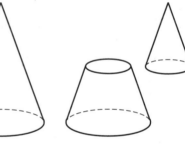

Exercise 19.4   Links: (*191*) 191

1 A toppling clown is made from a hemisphere and a cone.
   (a) Calculate the volume of the shape.
   (b) Calculate the surface area of the shape.

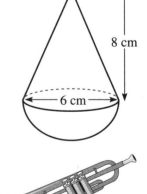

2 A trumpet mute is made in the shape of a truncated (cut-off) cone.
   (a) Calculate the volume of the trumpet mute.
   (b) Calculate the surface area of the trumpet mute.

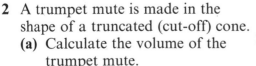

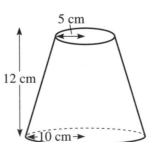

3 A model rocket is made from a cylinder and a cone. The radius
  of the cylinder and the cone is 5 cm. The cylinder has height
  20 cm and the cone has height 12 cm.
  (a) Calculate the volume of the shape.
  (b) Calculate the surface area of the shape.

20 Simplifying algebraic expressions

Exercise 20.1   **Links: (20A, B) 20A, B**

Simplify:

1 $(x^3)^2$

2 $(xy)^3$

3 $(x^2yz^3)^2$

4 $a^3 \times 3a$

5 $5a^2 \times 2a^3$

6 $a^2b^3 \times ab$

7 $abc \times ab$

8 $a^4 \times b^3c \times ab$

9 $(2x^2y)^3$

10 $2pq(p^2q + pq^2)$

11 $x^5 \div x^2$

12 $x^4 \div x$

13 $4x^6 \div 2x^3$

14 $a^2bc \div ab$

15 $15x^2z \div 3xz$

16 $9p^3q^4 \div 3p^2q^2$

17 $c^3 \div c$

Exercise 20.2   **Links: (20C, D, E) 20C, D, E**

Find the value of:

1 $27^{\frac{1}{3}}$

2 $\left(\frac{4}{9}\right)^{\frac{1}{2}}$

3 $(343)^{\frac{1}{3}}$

4 $(36)^{-\frac{1}{2}}$

5 $(121)^{-\frac{1}{2}}$

6 $(32)^{\frac{1}{5}}$

7 $(216)^{\frac{2}{3}}$

8 $9^{\frac{5}{2}}$

9 $(512)^{-\frac{2}{3}}$

10 $\left(\frac{9}{4}\right)^{-\frac{3}{2}}$

11 $\left(\frac{1}{16}\right)^{-\frac{1}{2}}$

12 $\left(\frac{36}{49}\right)^{\frac{3}{2}}$

13 16^0

14 $\left(\sqrt{8}\right)^{\frac{1}{3}}$

15 $(81)^{\frac{3}{4}} \div (9)^{\frac{1}{2}}$

Find the values of k:

16 $x^{2k} = (x^3)^{\frac{1}{2}}$

17 $x^k = 1 \div x^{-1}$

18 $y^{k+1} = (y^3)^3$

19 $\left(\sqrt{y}\right)^k = y$

20 $729^k = 9$

21 $16^k = 64$

22 $4^{\frac{k}{2}} = 32$

Simplify:

23 $(4a^2)^{\frac{1}{2}}$

24 $(27a^6b^3)^{\frac{1}{3}}$

25 $\left(\sqrt{2}x\right)^4$

26 $\left(\sqrt{3}y^2\right)^2$

27 $(a^{-2}b)^{-1}$

28 $(x^{-4}y^2)^{-\frac{1}{2}}$

29 $9x\sqrt{x}(3x)^{-\frac{3}{2}}$

30 $(16x^2y^{-2})^{-\frac{1}{2}}$

31 $(8x^6)^{\frac{1}{3}} + (4x^4)^{\frac{1}{2}}$

Exercise 20.3   Links: (*20F, G*) 20F, G

Simplify:

1 $\dfrac{4xy}{2y}$

2 $\dfrac{9x^2}{3x}$

3 $\dfrac{18xy^2}{3xy}$

4 $\dfrac{24a^2bc}{9abc^2}$

5 $\dfrac{2(x+2)}{y(x+2)}$

6 $\dfrac{2b^2}{a} \times \dfrac{a^3}{4b}$

7 $\dfrac{(x+1)(x-1)}{2(x-1)}$

8 $\dfrac{3x^2}{6(x+1)}$

9 $\dfrac{(z-2)^3}{(z-2)(z+2)}$

Write as a single fraction:

10 $\dfrac{1}{4}+\dfrac{2}{7}$

11 $\dfrac{x}{4}+\dfrac{2x}{7}$

12 $\dfrac{2x}{9}+\dfrac{x}{3}$

13 $\dfrac{1}{4}+\dfrac{1}{2x}$

14 $\dfrac{a}{3}+\dfrac{b}{5}$

15 $\dfrac{2c}{7}-\dfrac{3b}{5}$

16 $\dfrac{p+2}{3}+\dfrac{p-1}{5}$

17 $\dfrac{2q+3}{2}+\dfrac{3q-4}{3}$

18 $\dfrac{2}{y}-\dfrac{3}{4y}$

19 $\dfrac{1}{x+1}+\dfrac{1}{x+2}$

20 $\dfrac{2}{p-3}-\dfrac{5}{p+1}$

21 $\dfrac{a}{a+3}-\dfrac{3}{a+1}$

Write down the LCD for these denominators:

22 15, 40

23 74, 111

24 x^3, x^4

25 $xz, 2x^2$

26 $3(x+1), 5(x-1)$

Work out as a single fraction in its lowest terms:

27 $\dfrac{b+1}{bc}+\dfrac{1-a}{ac}$

28 $\dfrac{1}{a(b+1)}-\dfrac{1}{b(a+1)}$

29 $\dfrac{2x}{y-1}-\dfrac{x}{y}$

30 $\dfrac{3x}{2x+1}+\dfrac{1}{3}$

31 $\dfrac{x-1}{x+1}+\dfrac{x+1}{x-1}$

32 $\dfrac{1}{(x-1)^2}-\dfrac{1}{(x-1)}$

Exercise 20.4   Links: (*20H, I*) 20H, I

Factorize completely:

1 $6+2x$

2 $3a+3ab$

3 a^2+ac

4 a^3+2a^2b

5 $xy-x$

6 $15x-3x^2$

7 $y(x+y)+a(x+y)$

8 $b(2x-y)-c(2x-y)$

9 $(2x+1)(a+b)+(1-x)(a+b)$

10 $3c(2a+b)-d(b+2a)$

11 $2p(3 - q) + 3(q - 3)$

12 $ax + ay + bx + by$

13 $x^2 + xy + y^2 + yx$

14 $2pqr + 2pa - qr - a$

15 $x^2 + 3x - 15x - 45$

16 $x^2 + 4x + 8x + 32$

17 $x^2 - 24 - 6x + 4x$

18 $x^2 - 11x + 18$

19 $x^2 - 9x + 18$

20 $x^2 + 19x + 18$

21 $x^2 - 5x - 24$

22 $x^2 + 10x - 24$

23 $x^2 + 19x + 60$

24 $x^2 + 17x + 60$

25 $x^2 + 2x - 15$

26 $2x^2 - 7x - 15$

27 $3x^2 - x - 4$

28 $6x^2 - 19x - 7$

29 $12x^2 - 7x + 1$

30 $12x^2 - x - 6$

31 $10x^2 - 91x + 9$

32 $5x^2 + 44x - 9$

33 $5x^2 - 12x - 9$

34 $9 - 21x + 10x^2$

35 $x^4 + 3x^2 + 2$

36 $x^{2n} - 5x^n + 6$

Exercise 20.5   Links: (*20J, K*) 20J, K

Factorize completely:

1 $a^2 - 16$

2 $4y^2 - 81$

3 $36 - 9k^2$

4 $2x^2 - 50$

5 $5y^2 - 80$

6 $8z^2 - 162$

7 $x^2 + 16x + 64$

8 $x^2 - 8x + 16$

9 $2x^2 + 20x + 50$

10 $3x^2 - 36x + 108$

Simplify:

11 $x^2 - (x - 1)^2$

12 $(x + 2)^2 - (x + 1)^2$

13 $101^2 - 99^2$

14 $(2y + z)^2 - (2y - z)^2$

Simplify as fully as possible:

15 $\dfrac{x - 2}{x^2 - 4}$

16 $\dfrac{3x + 9}{x^2 - 9}$

17 $\dfrac{2x + 8}{x - 3} \times \dfrac{3x - 9}{x + 4}$

18 $\dfrac{x^2 - 49}{2x + 5} \div \dfrac{4x + 28}{4x^2 - 25}$

19 $\dfrac{1}{3x - 2} - \dfrac{x + 1}{3x^2 + 16x - 12}$

20 $\dfrac{x^2 - 5x + 6}{x^2 - 4} \times \dfrac{x^2 + 2x - 3}{x^2 - 9}$

21 $\dfrac{6x^2 + 13x + 5}{4x^2 - 4x - 3}$

22 $\dfrac{x^2 - 7x + 6}{x^2 - 5x - 6}$

21 Quadratics

Solve the equations:

1 $(x-4)(x-7)=0$ **2** $(x-2)(x+3)=0$

3 $(2x-1)(3x+4)=0$ **4** $3(1-3x)(3+2x)=0$

5 $(4x-1)(5x-2)=0$

Factorize and solve the quadratic equations:

6 $a^2-9a=0$ **7** $a^2+2a+1=0$

8 $b^2-7b+6=0$ **9** $b^2-5b=6$

10 $6c^2+11c-10=0$ **11** $2x^2-6x=0$

12 $(2x-1)^2=4$ **13** $(2-3x)^2-(x+1)^2$

14 $(x-3)^2+5(x-3)+4=0$ Hint: use $(x-3)=y$

1 Write the following in the form $(x+q)^2+r$.
   (a) x^2+2x **(b)** x^2+5x **(c)** x^2-8x

2 Write the following in the form $p(x+q)^2+r$.
   (a) $2x^2+6x$ **(b)** $3x^2-18x+3$ **(c)** $5x^2+40x-15$

3 Solve to 2 d.p. the equations:
   (a) $x^2+7x+9=0$ **(b)** $x^2-3x-8=0$
   (c) $2x^2-11x+6=0$ **(d)** $3x^2+8x-1=0$
   (e) $4x^2-15x+5=0$ **(f)** $3x^2=9-x$
   (g) $8=2x+11x^2$ **(h)** $7x=3x^2-5$

Solve to 2 d.p.:

1 $\dfrac{2}{x+1}-\dfrac{1}{x+2}=1$ **2** $\dfrac{4}{2x-5}-\dfrac{3}{x+7}=2$

3 $\dfrac{1}{x}-\dfrac{1}{x-1}=-3$ **4** $\dfrac{2}{(x-1)(2x-1)}+\dfrac{3x}{(x-1)}=2$

5 $\dfrac{(x+1)(3x-7)}{2x(x-3)}=1$ **6** $\dfrac{(x+1)}{(x+2)}=\dfrac{(2x+3)}{(x+1)}$

7 The area of this square is 40.
   (a) Form an equation in x.
   (b) Solve your equation.

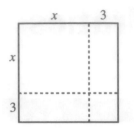

8 $h = ut - 5t^2$ is a formula which links height in metres (h) with
   initial velocity in m/s (u) and time in seconds (t).
   (a) If the initial velocity is $25\,\text{m/s}$ find the times when the height
       is 30 metres.
   (b) If the initial velocity is $18\,\text{m/s}$ find the times when the height
       is 16 metres.
   (c) Find the time if the initial velocity is $10\,\text{m/s}$ and the height is
       -15 metres.

9 The diagram is of a right-angled triangle with the length of the
   sides as shown.
   (a) Use Pythagoras' Theorem to form an equation in x.
   (b) Solve the equation to find x.

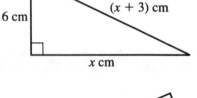

10 The diagram shows a tent in the shape of a triangular prism.
   The tent is 10 cm wider than it is high. It is 210 cm long and has
   a volume of 1386 litres.
   (a) Work out an expression for the cross-sectional area of the
       tent in terms of x.
   (b) Use the volume of 1386 litres to form an equation.
   (c) Find the height.

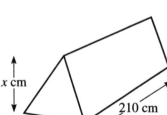

11 The diagram shows a circle with centre O.

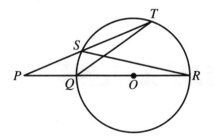

   Because triangles PQT and PSR are mathematically similar:

   $$PS \times PT = PQ \times PR$$

   (a) Given $PQ = 5$, $PS = 6$ and $ST = 3$ and the radius is x, form
       an equation in x and solve it.
   (b) Given that the radius is 9, $PS = 7$ and $ST = 2$, find PQ.

Exercise 21.4   Links: (*21G*) 21G, H

1 Solve the simultaneous equations:
 (a) $y = 18$ (b) $y = 4x$ (c) $y = 2x - 1$ (d) $y = 2x + 7$
     $y = 2x^2$ $y = x^2$ $y = x^2$ $y = x^2 - 4x$

2 Find the coordinates of the points of intersection:
 (a) $y + 5x = 12$ (b) $y = 3x^2 - 2x$
     $y = 2x^2$ $5y = 3x - 2$
 (c) $y = x^2 + 2x$ (d) $y = x^2 - 1$
     $7y = 6 - 5x$ $12y = 7x$

3 In each part solve the simultaneous equations and interpret your solution geometrically.
 (a) $x = 2$, $x^2 + y^2 = 16$
 (b) $5y = x + 39$, $x^2 + y^2 = 65$
 (c) $x^2 + y^2 = 65$, $x = 3y + 5$.

4 (a) Show that $y = 2x + 5$ is a tangent to the circle $x^2 + y^2 = 5$.
 (b) Find the equations of the tangents to the circle $x^2 + y^2 = 5$ that are perpendicular to $y = 2x + 5$.

5 Find the equation of the tangent to the circle $x^2 + y^2 = 50$ which touches the circle at $(7, 7)$.
 (Hint: the tangent to a circle is perpendicular to the radius.)

Exercise 21.5   Links: (*21G*) 21I

Solve these inequalities:

1 $x^2 \geqslant 64$ 2 $2x^2 < 98$ 3 $(2x)^2 \geqslant 100$

4 $4x^2 + 1 < 37$ 5 $3 - 2x^2 \geqslant -69$ 6 $9x^2 + 3 < 15.25$

7 $4(x^2 + 1) > 85$ 8 $(x + 3)^2 \geqslant 9$ 9 $(5 - x)^2 < 36$

10 $(2x + 1)^2 \geqslant 6.25$ 11 $\frac{1}{4}(2 - x)^2 \leqslant 9$ 12 $5(3 - x)^2 > 180x^2$

Exercise 21.6   Links: (*21H*) 21J

1 (a) On graph paper draw the graph of $y = 6 + 2x - x^2$ for values of x from -3 to $+4$.

 (b) On the same axes draw the graph of $y = \dfrac{7}{x}$ for $-3 \leqslant x \leqslant -\frac{1}{2}$ and $\frac{1}{2} \leqslant x \leqslant 4$

 (c) Use your graphs to find approximate solutions to
 (i) $6 + 2x - x^2 = 0$ (ii) $2x - x^2 = -4$ (iii) $6 + 2x - x^2 = \dfrac{7}{x}$

2 **(a)** On graph paper draw the graph of $y = x^3 - 4x + 4$ for
$-3 \leqslant x \leqslant 3$

(b) On the same axes draw the graph of $y = 2x + 8$

(c) Use your graphs to find approximations for

(i) $x^3 - 4x + 4 = 4$ **(ii)** $x^3 - 4x = -8$

(iii) $x^3 - 4x = 2x + 4$ **(iv)** $2x^3 - 8x + 8 = 3$

3 **(a)** On graph paper draw the graph of $y = 9x - x^3$ for
$-3 \leqslant x \leqslant 3$

(b) On the same axes draw the graph of $y = x - \dfrac{1}{x}$ for

$-3 \leqslant x \leqslant -0.1$ and $0.1 \leqslant x \leqslant 3$

(c) Use your graphs to find approximate solutions to

(i) $x - \dfrac{1}{x} = 0$ **(ii)** $x^3 - 9x = 0$

(iii) $x^3 = 9x$ **(iv)** $9x - x^3 = x - \dfrac{1}{x}$

(v) $\dfrac{1}{x} + 8x - x^3 = 0$ **(vi)** $\dfrac{1}{x} - x = 2$

22 Advanced trigonometry

1 Calculate the area of each of the triangles to 2 d.p.

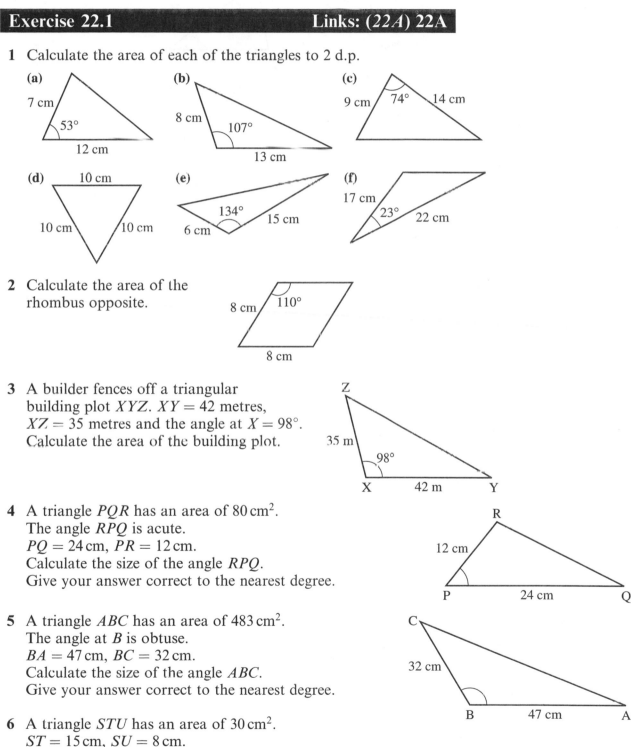

(a) 7 cm, 53°, 12 cm

(b) 8 cm, 107°, 13 cm

(c) 9 cm, 74°, 14 cm

(d) 10 cm, 10 cm, 10 cm

(e) 134°, 15 cm, 6 cm

(f) 17 cm, 23°, 22 cm

2 Calculate the area of the rhombus opposite.

8 cm, 110°, 8 cm

3 A builder fences off a triangular building plot XYZ. $XY = 42$ metres, $XZ = 35$ metres and the angle at $X = 98°$. Calculate the area of the building plot.

Z, 35 m, 98°, X, 42 m, Y

4 A triangle PQR has an area of $80\,\text{cm}^2$.
The angle RPQ is acute.
$PQ = 24\,\text{cm}$, $PR = 12\,\text{cm}$.
Calculate the size of the angle RPQ.
Give your answer correct to the nearest degree.

R, 12 cm, P, 24 cm, Q

5 A triangle ABC has an area of $483\,\text{cm}^2$.
The angle at B is obtuse.
$BA = 47\,\text{cm}$, $BC = 32\,\text{cm}$.
Calculate the size of the angle ABC.
Give your answer correct to the nearest degree.

C, 32 cm, B, 47 cm, A

6 A triangle STU has an area of $30\,\text{cm}^2$.
$ST = 15\,\text{cm}$, $SU = 8\,\text{cm}$.
Calculate both of the possible sizes of the angle TSU.

Exercise 22.2   Links: (*22B, C, D*) 22B, C, D

1 Work out the sides or angles marked with a letter.

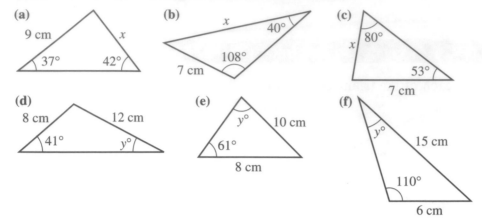

(a)
9 cm   *x*
37°   42°

(b)
x   40°
108°
7 cm

(c)
x   80°
53°
7 cm

(d)
8 cm   12 cm
41°   *y*°

(e)
y°   10 cm
61°
8 cm

(f)
y°
15 cm
110°
6 cm

2 A lighthouse, *L*, lies 52 km due north of a marker buoy, *B*. A trawler, *T* lies on a bearing of 035° from *B* and on a bearing of 048° from *L*. Work out the distance from *L* to *T*.

3 The area of a triangle *ABC* can be written as $\frac{1}{2}ab\sin C$.
   (a) Write the area in two other ways, using angles *A* and *B* respectively.
   (b) By equating the expressions for the area of *ABC*, establish the **sine rule**.

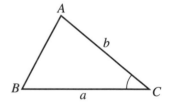

4 *PQR* is a triangular plot of land.
   PQ = 32 m, *PR* = 37 m and
   the angle *PQR* = 45°.
   Work out the angle *PRQ*.
   Give your answer to the nearest degree.

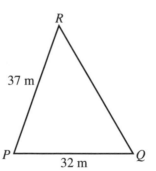

R
37 m
P   32 m   *Q*

5 The diagram shows the relative positions of three villages, Ashwell (*A*), Bredbury (*B*) and Crimpton (*C*).
   The bearing of Bredbury from Ashwell is 023°
   The bearing of Crimpton from Ashwell is 075°
   The bearing of Crimpton from Bredbury is 114°
   (a) Show that the angle *ACB* = 39°.

   The distance from Ashwell to Bredbury is 8.6 km.
   (b) Work out the distance from Bredbury to Crimpton.

N
Bredbury
8.6 km
23°
Crimpton
75°
Ashwell

6 A woman leaves her home, *H*, and walks 1.2 km due east to reach a barn, *B*. At *B*, she turns through an angle of 130° and continues to walk in a straight line until she reaches a shed *S*. The angle *BSH* is 27°.

Calculate how far the woman would have walked if she had walked in a straight line from her home to the shed.

7 In a triangle *XYZ*, *XY* = 12 cm, *YZ* = 7 cm and the angle at *X* = 23°.

(a) Show that there are two possible values for the angle at *Z*.

(b) Calculate each of the values of the angle at *Z*.

Exercise 22.3   Links: (22E) 22E

1 Calculate each of the sides or angles marked with a letter.

(a)
5 cm, 58°, 8 cm, *x*

(b)
x, 142°, 12 cm, 9 cm

(c)
9 cm, 48°, *x*, 14 cm

(d)
7 cm, 4 cm, *y*°, 9 cm

(e)
12 cm, 6 cm, *y*°, 8 cm

(f)
9 cm, 5 cm, *y*°, 12 cm

2 A ship leaves a port, *P* and travels 46 km due north to reach a lighthouse, *L*. At *L* the ship turns on to a bearing of 310° and travels a further 71 km to reach a marker buoy, *B*. At *B* the ship turns again and travels in a straight line back to *P*.

Calculate the total distance travelled by the ship.

3 The lengths of the sides of a triangle are 5 cm, 12 cm and 15 cm. Work out the three angles of this triangle.

Give your answers correct to the nearest degree.

4 In a triangle *ABC*, the lengths of the sides, in cm, are

    AB = *c*   *BC* = *a*   and   *AC* = *b*

The angle *ACB* = 120°

Show that

$$c^2 = a^2 + b^2 + ab$$

5 *PQRS* is a parallelogram with *PQ* = 12 cm, *QR* = 7 cm and the angle *PQR* = 150°. Work out the lengths of the two diagonals of the parallelogram.

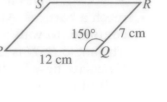

6 The perimeter of a triangle is 50 cm.
The length of one of the sides is 24 cm, the length of another of the sides is 23 cm. Calculate, to the nearest degree, the size of the smallest angle of this triangle.

7 A port, *P*, is 23 km due north of a harbour, *H*.
At 12:00 a yacht sets out from *P* and travels at 12 km/h on a bearing of 072°.
Also at 12:00 a ship sets out from *H* and travels on a straight course in a general north easterly direction.
The yacht and the ship meet up at 14:00 hours.
Calculate:
(a) the speed of the ship
(b) the bearing on which the ship travelled

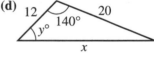

Exercise 22.4   Links: (*22F*) 22F

1 For each of these triangles, calculate:
   (i) the area   **(ii)** the length of the side marked *x*
   (iii) the angle marked *y*°.   (Note: all lengths are in cm.)

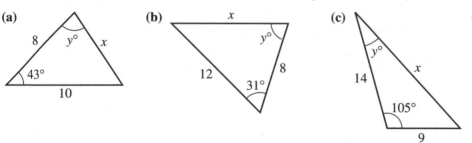

(a) **(b)** **(c)** **(d)**

2 A ship leaves a harbour, *H*, and travels 38 km due north to a marker buoy, *B*. At *B* the ship turns on a bearing of 048° and travels for a further 27 km to reach a lighthouse, *L*.
At *L* the ship turns again and travels in a straight line back to *H*.
Calculate:
(a) the total distance travelled by the ship
(b) the bearing of *L* from *H*
(c) the shortest distance between the ship and *B* on the ship's return journey to *H*.

3 In triangle *ABC*, *AB* = 5 cm,
AC = *x* cm, *BC* = 2*x* cm and
angle *BAC* = 60°.
Show that $3x^2 + 5x - 25 = 0$ [L]

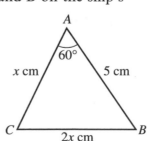

4 Two cars set off from the same point at 12:00. The first car travels at a constant speed of 65 km/h on a bearing of 043°. The second car travels at a constant speed of 74 km/h on a bearing of 290°.
Calculate the distance between the two cars at precisely 14:00 hours.

5 The diagram represents a children's slide which stands on level ground. *SR* is a straight line on the ground and *T* is the top of the slide. *ST* = 8 m, *RT* = 12 m and the angle *RST* = 43°.
Calculate:
(a) the height of *T* above the ground
(b) the angle *SRT*
(c) the distance between *S* and *R*

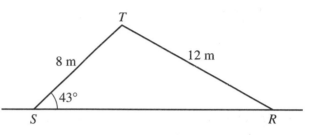

6 The lengths of the sides of a triangle are 6 cm, 8 cm and 11 cm. Calculate the area of the triangle.

Exercise 22.5   Links: 22G

1 *VABC* is a solid tetrahedron.
The vertex *V* is vertically above *B* with *VB* = 24 cm.
The horizontal base is a triangle *ABC* with the angle at *B* being 90°.
AB = 10 cm   and   *BC* = 7.5 cm
(a) Calculate the lengths of:
   (i) *AC*   (ii) *VC*   (iii) *VA*
(b) Calculate the angles:
   (i) *BVC*   (ii) *VAB*   (iii) *AVC*
(c) Calculate the surface area of *ABCV*.

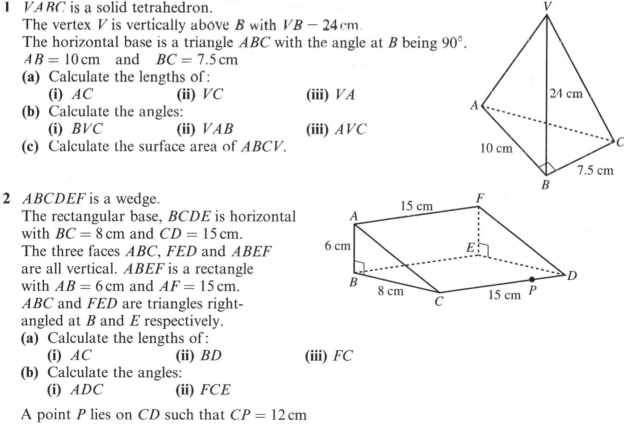

2 *ABCDEF* is a wedge.
The rectangular base, *BCDE* is horizontal with *BC* = 8 cm and *CD* = 15 cm.
The three faces *ABC*, *FED* and *ABEF* are all vertical. *ABEF* is a rectangle with *AB* = 6 cm and *AF* = 15 cm.
ABC and *FED* are triangles right-angled at *B* and *E* respectively.
(a) Calculate the lengths of:
   (i) *AC*   (ii) *BD*   (iii) *FC*
(b) Calculate the angles:
   (i) *ADC*   (ii) *FCE*

A point *P* lies on *CD* such that *CP* = 12 cm
(c) Calculate the angles:
   (i) *FPE*   (ii) *EPB*

3 *VABCD* is a rectangular-based pyramid.
The vertex *V* is vertically above *M*, the midpoint of the
horizontal base *ABCD*.
$AB = 16$ cm, $AC = 34$ cm and $VA = 25$ cm.
(a) Calculate the lengths of:
    (i) *BC*   (ii) *VM*
(b) Calculate the angles:
    (i) *VBM*   (ii) *BVA*   (iii) *AVD*

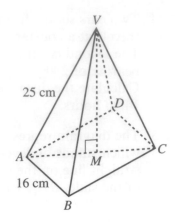

4 *PQRS* is a regular tetrahedron with
each side of length 10 cm.
It is placed so that the face *PQR* is
on a horizontal table.
Work out the height of *S* above
the table.

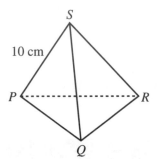

5 The diagram shows a tetrahedron *VABC*.
The horizontal base is a triangle *ABC* with
$AB = 16$ cm, $BC = 24$ cm and the angle $ABC = 120°$.

The vertex *V* is vertically above *B* and $VB = 30$ cm
(a) Calculate the area of the base *ABC*.
(b) Calculate the lengths of
    (i) *VA*   (ii) *VC*   (iii) *AC*
(c) Calculate the angles
    (i) *AVB*   (ii) *VCB*   (iii) *AVC*
    (iv) *ACV*   (v) *BAC*

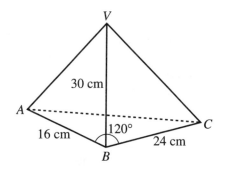

6 *VABCD* is a pyramid.
The horizontal base *ABCD* is a rhombus of side 8 cm.
The angle $ABC = 135°$
(a) Calculate the length of each diagonal of the base of the
    pyramid.

The vertex *V* is vertically above the centre of the base.
$VA = 20$ cm.
(b) Calculate the height of *V* above the base.

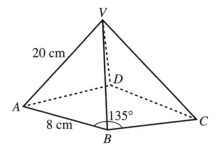

23 Exploring numbers

1 Convert the following fractions into decimals and indicate which of the decimals are terminating and which are recurring:

$$\frac{2}{3}, \quad \frac{1}{4}, \quad \frac{4}{5}, \quad \frac{5}{6}, \quad \frac{9}{13}$$

2 Write out the recurring decimals which are equivalent to the fractions:

$$\frac{1}{9}, \quad \frac{2}{9}, \quad \frac{3}{9}, \quad \frac{4}{9}, \quad \frac{5}{9}, \quad \frac{6}{9}, \quad \frac{7}{9}, \quad \frac{8}{9}$$

Explain the pattern of recurring decimals.

3 Find the decimals equivalent to the fractions:

$$\frac{1}{15}, \quad \frac{2}{15}, \quad \frac{3}{15}, \quad \cdots \quad \frac{12}{15}, \quad \frac{13}{15}, \quad \frac{14}{15}$$

(a) Separate the decimals into two distinct sets.
(b) Explain the relationship between the two sets.

4 Find fractions which are equivalent to the recurring decimals:
(a) $0.3333\ldots$ (b) $0.8888\ldots$
(c) $0.575757\ldots$ (d) $0.01230123\ldots$
(e) $5.232323\ldots$ (f) $3.8565656\ldots$

5 Find the fraction which is equivalent to the recurring decimal $0.0999999\ldots$

1 Solve $x^2 = 40$, leaving your answer in its most simplified surd form.

2 The area of a square is $90\,\text{cm}^2$. Find the length of one side of the square. Give your answer in its most simplified surd form.

3 Simplify:

(a) $\frac{1}{\sqrt{5}}$ (b) $\frac{2}{\sqrt{7}}$ (c) $\frac{\sqrt{6}}{\sqrt{7}}$

(d) $\frac{1}{\sqrt{13}}$ (e) $\frac{\sqrt{128}}{\sqrt{2}}$ (f) $\frac{\sqrt{11}}{\sqrt{22}}$

4 A rectangle has sides of length $5 + \sqrt{3}$ and $5 - \sqrt{3}$ units. Work out, in their most simplified form:
(a) the perimeter of the square
(b) the area of the square
(c) the length of a diagonal of the rectangle.

5 Solve the equation:

    (a) $x^2 - 10x - 32 = 0$ **(b)** $x^2 + 6x + 3 = 0$

6 Show that: $\frac{1}{3\sqrt{13}} = \frac{\sqrt{13}}{39}$

Exercise 23.3   Links: (*23D, E, F*) 23D, E, F

1 Write the greatest lower bound and least upper bound for these measurements, to the given degree of accuracy.

    (a) To the nearest unit:

        5,   15,   105,   115,   1005

    (b) To the nearest 1000:

        15 000,   1000,   9000,   5000,   10 000

    (c) To the nearest 0.5 unit:

        3.5,   11,   17.5,   21,   22.5

    (d) To the nearest 0.25 unit:

        11,   3.25,   9.75,   50,   17.5

2 The 400 m record at a running track in Stevenage is 44.152 seconds correct to the nearest $\frac{1}{1000}$ of a second.

    (a) Write down the least upper bound and greatest lower bound for this record.

    (b) A runner records a lap time of 44.1516 seconds. Explain whether this is a new record.

3 Using the least upper bound and greatest lower bound for these measurements, find the least upper and greatest lower bounds for these quantities. The degree of accuracy of each measurement is given. Use the π button on your calculator when needed.

    (a) The area of a rectangle with sides 5 cm and 9 cm both measured to the nearest cm.

    (b) The perimeter of a rectangle with sides 5 cm and 9 cm both measured to the nearest cm.

    (c) The area of a circle with radius 4.2 cm measured to 1 d.p.

    (d) The circumference of a circle with radius 5.19 cm measured to 2 d.p.

    (e) The area of a triangle with base length 6.1 cm and perpendicular height 12.5 cm, both measured to the nearest mm.

    (f) The area of a triangle in which two sides are of length 8 cm and 12 cm and the size of the angle between them is 62°, all measured to the nearest unit.

    (g) The volume of a cone with:

        (i) radius 25 cm and height 40 cm both measured to the nearest 5 cm

       (ii) radius 25 cm and height 40 cm both measured to 2 s.f.

     (iii) radius 25.0 cm and height 40.0 cm both measured to 1 d.p.

4 Calculate the least upper bound and greatest lower bound for the difference in length of two pieces of string. One measures 160 cm, the other 82 cm, both measured to 2 s.f.

5 A car travels 90 km in 1.5 hours, where both measurements are correct to 2 s.f. Calculate the least upper bound and greatest lower bound of the car's average speed in km/h.

6 The diagram shows triangle *ABC* where angle $ACB = 105°$ to 3 s.f., $AC = 18$ cm to 2 s.f. and $BC = 25$ cm to 2 s.f. Use the cosine rule to calculate the greatest lower bound and the least upper bound of the length *AB*.

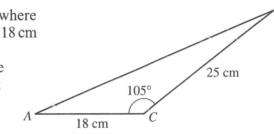

7 A sphere has a radius of 0.45 cm correct to 2 d.p. Calculate the least upper bound and greatest lower bound of the volume and surface area of the sphere.

8 $x = 20$ and $y = 30$, correct to the nearest 5
Work out the minimum value of

$$\frac{x + y}{x}$$

9 $x = 4$ and $y = 0.07$, correct to 1 significant figure.

$$z = y - \frac{x - 8}{x}$$

Work out the least upper bound for *z*.

Exercise 23.4   Links: (*23G, H*) 23G, H

1 A bag of icing sugar is labelled as 500 g. The actual weight is found to be 525 g. Calculate the absolute error and the percentage error of the weight.

2 Joe estimated the area of a circle with radius 55 cm using the approximation

$$3 \times 60 \times 60 = 10\,800 \text{ cm}^2$$

Calculate the absolute and percentage error of this estimate. (Use the π button on your calculator).

3 In a magazine the length of a bookcase is given as 120 cm correct to 2 s.f. Calculate the maximum possible absolute error and the maximum percentage error of the length of the bookcase.

4 The width of a rectangle was measured as 15 cm and the length as 28 cm both correct to 2 s.f. Calculate the maximum absolute error and maximum percentage error of:
(a) the length and width of the rectangle
(b) the area of the rectangle
(c) the perimeter of the rectangle

5 A triangle has base length 4.6 cm and perpendicular height 8.9 cm both correct to 1 d.p. Calculate the least upper bound and the greatest lower bound of the area of the triangle.

6 $AB = 8.0$ cm and $BC = 12.4$ cm, correct to the nearest mm. Work out the maximum and minimum value of angle $\bigcirc{C}$.

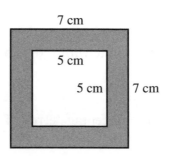

Exercise 23.5   Links: (*23I*) 23H

1 Write as fractions:
(a) $0.1\dot{2}\dot{7}$ (b) $0.1\dot{3}4\dot{3}4$

2 Simplify:
(a) $\frac{2}{\sqrt{3}}$ (b) $\frac{\sqrt{2}}{\sqrt{3}}$ (c) $\frac{1}{\sqrt{15}}$

3 Solve:

$$x^2 - 8x + 11 = 0$$

Give your answer in surd form.

4 Write the least upper bound and greatest lower bound for:
(a) 450 000 to 3 s.f.   (b) 7.1×10^{-5} to 2 s.f.

5 The squares have sides 7 cm and 5 cm each correct to 1 s.f. Find the least upper bound and greatest lower bound of the shaded area. State the shaded area to an appropriate degree of accuracy.

6 The length of the side of a cube is 3.4 cm correct to 2 s.f. Calculate the maximum absolute error and maximum percentage error for
(a) the side of the cube   (b) the surface area of the cube
(c) the volume of the cube.

24 Applying transformations to sketch graphs

1 $f(x) = 3x + 1$, find:
   (a) $f(2)$ (b) $f(7)$ (c) $f(0)$ (d) $f(-4)$ (e) $f\left(\frac{1}{4}\right)$

2 $f(x) = 3x$, find:
   (a) $f(-x)$ (b) $f(3x)$ (c) $f(x+1)$
   (d) $f(2x+1)$ (e) $-f(x)$ (f) $-f(-x)$

3 $f(x) = x^3 - 1$, find:
   (a) $f(0)$ (b) $f(-1)$
   (c) $f(3)$ (d) $f\left(\frac{1}{2}\right)$
   (e) $f(2x)$ (f) $f(-x)$
   (g) $\frac{1}{2}f(x)$ (h) $f(1-x)$

Copy and complete:

1 (a) The graph of $y = 2x + 3$ is the graph of $y = 2x$ translated ...
   (b) The graph of $y = 3x^2 - 2$ is the graph of $y = 3x^2$
       translated ...
   (c) The graph of $y = \dfrac{1}{x} + 4$ is the graph of $y = \dfrac{1}{x}$ ...
   (d) The graph of $y = 3x + 4$ is the graph of ... translated 4 units
       vertically in the positive y-direction.
   (e) The graph of $y = x^2 + 5$ is the graph of $y = \ldots$ translated 3
       units vertically in the positive y-direction.
   (f) The graph of $y = x^3 + 1$ is the graph of ...

1 Show how the following graphs are related to the graph of
   $y = x^2$.
   In each case give the coordinates of the vertex.
   (a) $y = x^2 + 2x + 1$ (b) $y = x^2 - 2x + 1$
   (c) $y = x^2 + 16x + 64$ (d) $y = x^2 + 5x + 6.25$

2 Copy and complete:

(a) $y = (x - 2)^2$ is the graph of $y = x^2$ translated ... units in the positive x-direction.

(b) $y = (x + 3)^3$ is the graph of $y = x^3$ translated ...

(c) $y = \dfrac{1}{(x + 2)}$ is the graph of $y = \dfrac{1}{x}$ ...

(d) $y = 3(x + 1)$ is the graph of $y = 3x$ ...

3 (a) $y = (x - 3)^3 + 1$ is the graph of $y = x^3$ translated 3 units ... and 1 unit ...

(b) $y = \dfrac{1}{(x - a)} + b$ is the graph of $y = \dfrac{1}{x}$ translated ...

(c) $y = (x + 2)^3 - c$ is the graph of $y = x^3$ translated ...

(d) $y = 2x + 5 = 2(x + 2) + 1$ is the graph of $y = 2x$ translated ...
$ = 2(x + 1) + 3$ is the graph of $y = 2x$ translated ...
$ = 2(x) + 5$ is the graph of $y = 2x$ translated ...

4 Sketch the graph of $y = x^2$.

(a) Use a tracing of this to sketch on the same axes the graph of
$y = (x - 3)^2 + 1 = x^2 - 6x + 10$.

(b) Rearrange the equation $y = x^2 - 6x + 5$ as $y = (x - p)^2 + q$ in order to sketch the graph.

5 Relate the following graphs to $y = x^2$ using the method of question **4(b)**.

In each case give the coordinates of the vertex.

(a) $y = x^2 + 6x + 7$

(b) $y = x^2 - 2x + 8$

(c) $y = x^2 + 5x + 0.25$

6 Sketch the curve $y = \dfrac{1}{x}$ for $x > 0$.

(a) Use a tracing of this to sketch on the same axes:

$$y = \dfrac{1}{(x - 1)} + 2 = \dfrac{2x - 1}{x - 1}$$

(b) Sketch $y = \dfrac{1}{x + 3} - 1$

Exercise 24.4   Links: (*24G, H, I*) 24G, H, I

1 This is the graph of $y = f(x)$ where

$f(x) = (x - 1)(x - 3)$

$\quad = x^2 - 4x + 3$

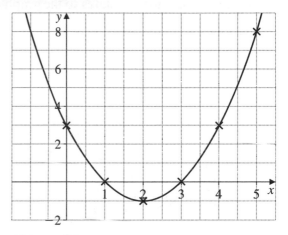

(a) Find the equation of $f(-x)$.
(b) Sketch the graph of $f(-x)$
(c) Sketch the graph of $-f(x)$ and write down the equation.

Mark the coordinates of all intercepts and vertices.

2 Make a flow chart which transforms $y = f(x)$, where $f(x) = x^2$ into

$\qquad y = -x^2 - 4x + 3$

via $\quad y = -(x^2 + 4x) + 3$

$\qquad y = -((x + 2)^2 - 4) + 3$

$\qquad y = -(x + 2)^2 + 7$

$\qquad y = -f(x + 2) + 7$

List the separate transformations and draw them on a graph.

3 Draw the flow chart which transforms $y = \dfrac{1}{x}$ into $y = \dfrac{-1}{x - 3} + 2$

List the separate transformations in order.

4 Draw the flow chart which transforms $y = x^3$ into $y = x^3 - 3x^2 + 3x - 7$.
List the separate transformations in order.

5 List the transformations, in the correct order, which when applied to $y = x^2$ give the graphs of the following:

(a) $y = x^2 + 6x - 2$ **(b)** $y = 4x - x^2$

(c) $y = 2x^2 + 2x - \frac{1}{2}$ **(d)** $y = 4x^2 + 2$

(e) $y = 1 - 9x^2$

Sketch each graph showing all intercepts and vertices.

6 List the transformations, in the correct order, which when applied to $y = \dfrac{1}{x}$ give the graphs of the following:

(a) $y = \dfrac{1}{3x} - 1$ **(b)** $y = \dfrac{3}{x} - 1$

(c) $y = 2 - \dfrac{4}{2 + x}$ **(d)** $\dfrac{-1}{2(x - 2)}$

Sketch each graph.

7 For each of the graphs in the diagram, select which function represents
   them and say how it relates to the graph of the function sin x.

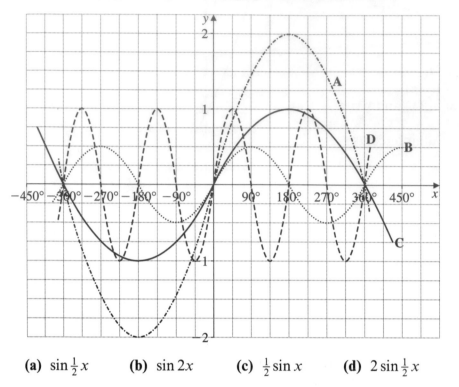

   (a) $\sin \frac{1}{2}x$ **(b)** $\sin 2x$ **(c)** $\frac{1}{2}\sin x$ **(d)** $2\sin \frac{1}{2}x$

8 The diagram shows $f(x) = \tan x$.
   Explain why $\tan x = \tan(x + 180°)$.

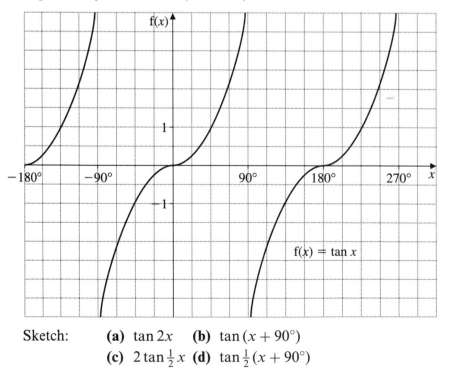

   Sketch: **(a)** $\tan 2x$ **(b)** $\tan(x + 90°)$
                (c) $2\tan \frac{1}{2}x$ **(d)** $\tan \frac{1}{2}(x + 90°)$

25 Vectors

1 P is the point (2, 1), Q is the point (4, 5) and R is the point (−2, 3). Write the column vectors:

   (a) $\overrightarrow{PQ}$ (b) $\overrightarrow{QR}$ (c) $\overrightarrow{RQ}$ (d) $\overrightarrow{PR}$

2 $\overrightarrow{CD}$ is the vector $\begin{pmatrix} 3 \\ -2 \end{pmatrix}$ and $\overrightarrow{DE}$ is the vector $\begin{pmatrix} 5 \\ -4 \end{pmatrix}$.

   Find the vectors: (a) $\overrightarrow{CE}$ (b) $\overrightarrow{EC}$

3 $a = \begin{pmatrix} 2 \\ 3 \end{pmatrix}$, $b = \begin{pmatrix} 5 \\ -8 \end{pmatrix}$, $c = \begin{pmatrix} -5 \\ -3 \end{pmatrix}$.

   (a) Write as column vectors:
(i) 2a	(ii) 3a + 2c	(iii) 5b − 3c	(iv) 4c + 3a
(v) 4a − 2b	(vi) 5b − 2a	(vii) −3b	(viii) b − a

   (b) Find the vector **d** such that:
      (i) $2a + d = c$ (ii) $d − d = 2c$

   (c) Write down a vector that is parallel to $a + b + c$.

4 $p = \begin{pmatrix} 5 \\ 2 \end{pmatrix}$, $q = \begin{pmatrix} -2 \\ 4 \end{pmatrix}$, $r = \begin{pmatrix} -1 \\ -1 \end{pmatrix}$.

   (a) Write as column vectors:
(i) 3p	(ii) 2p + 3r	(iii) 5q − 3r	(iv) 2r + 5p
(v) 2p − 2q	(vi) 7q − 2p	(vii) −6q	(viii) q − p

   (b) Find the vector **s** such that:
      (i) $3p + s = r$ (ii) $s − 3q = 5r$

   (c) Write down a vector that is parallel to $p + q + r$ and twice the length of it.

5 Find the magnitude of these vectors:

   (a) $\begin{pmatrix} 4 \\ 3 \end{pmatrix}$ (b) $\begin{pmatrix} -6 \\ 8 \end{pmatrix}$ (c) $\begin{pmatrix} -1 \\ -1 \end{pmatrix}$ (d) $\begin{pmatrix} 5 \\ 12 \end{pmatrix}$ (e) $\begin{pmatrix} 2 \\ -2 \end{pmatrix}$

6 $p = \begin{pmatrix} 3 \\ 4 \end{pmatrix}$, $q = \begin{pmatrix} -5 \\ 12 \end{pmatrix}$, $r = \begin{pmatrix} -1 \\ -1 \end{pmatrix}$.

   Work out:
(a) $\lvert p \rvert$	(b) $\lvert q \rvert$	(c) $\lvert r \rvert$	(d) $\lvert p + q \rvert$	(e) $\lvert p − q \rvert$
(f) $2\lvert r \rvert$	(g) $\lvert 3r \rvert$	(h) $\lvert p + r \rvert$	(i) $\lvert q − r \rvert$	(j) $\lvert p + q − r \rvert$

7 $\mathbf{a} = \begin{pmatrix} 2 \\ 3 \end{pmatrix}$, $\mathbf{b} = \begin{pmatrix} -2 \\ 1 \end{pmatrix}$, $\mathbf{c} = \begin{pmatrix} -1 \\ -1 \end{pmatrix}$.

Calculate **x** when:

(a) $\mathbf{a} + \mathbf{x} = \mathbf{b}$ (b) $2\mathbf{b} + \mathbf{x} = \mathbf{c}$

(c) $3\mathbf{c} + \mathbf{x} = 2\mathbf{a}$ (d) $\mathbf{x} - 4\mathbf{c} = \mathbf{a}$

(e) $3\mathbf{x} - \mathbf{a} = \mathbf{b}$

8 $\mathbf{p} = \begin{pmatrix} 2 \\ -1 \end{pmatrix}$, $\mathbf{q} = \begin{pmatrix} -2 \\ -3 \end{pmatrix}$, $\mathbf{r} = \begin{pmatrix} -1 \\ 0 \end{pmatrix}$.

(a) $2\mathbf{p} + a\mathbf{q}$ is parallel to the x-axis. Find the value of a.

(b) $3\mathbf{p} - b\mathbf{r}$ is parallel to the y-axis. Find the value of b.

9 $\mathbf{a} = \begin{pmatrix} 2 \\ 1 \end{pmatrix}$, $\mathbf{b} = \begin{pmatrix} -1 \\ 3 \end{pmatrix}$.

Find the values of p and q such that $p\mathbf{a} + q\mathbf{b} = \begin{pmatrix} 5 \\ -8 \end{pmatrix}$

10 A is the point $(3, 2)$, O is the point $(0, 0)$ and the position of a variable point P is given by:

$$\overrightarrow{OP} = \overrightarrow{OA} + t\begin{pmatrix} 1 \\ 1 \end{pmatrix}$$

Calculate the coordinates of P for integer values of t between -2 and $+4$. Write down the equation of the path of P, as t varies, in the form $y = mx + c$.

11 P is the point $(2, 1)$, Q is the point $(1, 3)$ and R is the point $(-2, -2)$.

(a)   Write down the coordinates of the midpoints of:

     (i) PQ **(ii)** QR **(iii)** PR

(b) S lies on PQ extended so that $PS = 3 \times PQ$. Work out the coordinates of S.

(c) T lies on PR so that $PT = \frac{1}{3}PR$. Work out the coordinates of T.

12 $ABCD$ is a rhombus. A has coordinates $(1, 2)$, B has coordinates $(3, 5)$ and the coordinates of the point of intersection of the diagonals is $(3, 2)$.

(a)   Find the position vectors of the points C and D.

(b)   Work out the vector of the line joining the midpoint of AB and CD.

1 $\overrightarrow{AB} = \mathbf{a}$ and $\overrightarrow{AC} = \mathbf{c}$. D and E are the midpoints of AB and AC respectively.
   (a) Write down in terms of $\mathbf{a}$ and $\mathbf{c}$ the vectors:
      (i) $\overrightarrow{AD}$ (ii) $\overrightarrow{AE}$ (iii) $\overrightarrow{DE}$ (iv) $\overrightarrow{BC}$
   (b) Write down the geometrical relationship between the lines DE and BC.

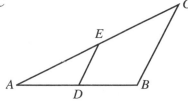

2 $\overrightarrow{AB} = \mathbf{b}$ and $\overrightarrow{AC} = \mathbf{c}$. D and E are $\frac{3}{4}$ of the way along AB and AC respectively. Find $\overrightarrow{BC}$ and $\overrightarrow{DE}$ in terms of $\mathbf{b}$ and $\mathbf{c}$ and write down the geometrical relationship between the lines DE and BC.

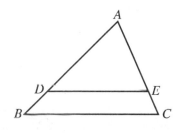

3 $\overrightarrow{AE} = \mathbf{e}$ and $\overrightarrow{AD} = \mathbf{d}$. $CE:AE = 2:1$ and $BD:AD = 2:1$.
   Write the vectors $\overrightarrow{DE}$ and $\overrightarrow{BC}$ in terms of $\mathbf{e}$ and $\mathbf{d}$ and then write down two facts about the lines DE and BC.

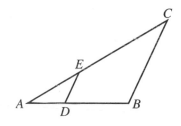

4 $ABCD$ is a parallelogram, P, Q, R and S are the midpoints of AB, BC, CD and DA respectively.
   $\overrightarrow{AB} = \mathbf{a}$ and $\overrightarrow{AD} = \mathbf{d}$.
   Write $\overrightarrow{PQ}$, $\overrightarrow{QR}$, $\overrightarrow{SR}$ and $\overrightarrow{PS}$ in terms of $\mathbf{a}$ and $\mathbf{d}$.
   Write down the name of the quadrilateral $PQRS$.

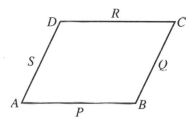

5 $\overrightarrow{AD} = \mathbf{x}$ and $\overrightarrow{AE} = \mathbf{y}$. $AC = 3AE$ and $AB = 3AD$.
   (a) (i) Find $\overrightarrow{DE}$ and $\overrightarrow{BC}$ in terms of $\mathbf{x}$ and $\mathbf{y}$.
      (ii) Hence explain why triangles PDE and PBC are similar.
   (b) Express $\overrightarrow{DP}$ and $\overrightarrow{AP}$ in terms of $\mathbf{x}$ and $\mathbf{y}$.

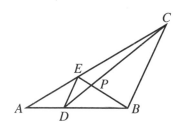

6 $ABCDEF$ is a regular hexagon with its centre at point O.
   $\overrightarrow{OB} = \mathbf{b}$, $\overrightarrow{OC} = \mathbf{c}$.
   (a) Write in terms of $\mathbf{b}$ and $\mathbf{c}$ the vectors:
      (i) $\overrightarrow{BC}$ (ii) $\overrightarrow{CD}$ (iii) $\overrightarrow{DE}$ (iv) $\overrightarrow{EF}$
      (v) $\overrightarrow{FA}$ (vi) $\overrightarrow{AB}$ (vii) $\overrightarrow{AC}$ (viii) $\overrightarrow{FD}$
   (b) Prove that $ACDF$ is a rectangle.

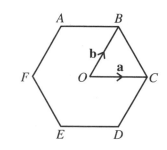

7 In triangle ABC, M is the midpoint of BC and N is the midpoint of AC. $\overrightarrow{AB} = \mathbf{b}$ and $\overrightarrow{BC} = \mathbf{c}$.
BN is produced to a point D so that $BD:BN = 4:3$.
   (a) Write $\overrightarrow{AM}$ and $\overrightarrow{DC}$ in terms of $\mathbf{b}$ and $\mathbf{c}$.
   (b) Prove that AM is parallel to DC and $AM:DC = 3:2$.

8 In triangle ABC, P is the point on AB so that $AP = \frac{2}{3}AB$ and Q is the midpoint of BC. AC is produced to a point R so that $AC = CR$. $\overrightarrow{AB} = \mathbf{b}$ and $\overrightarrow{AC} = \mathbf{c}$.
   (a) State in terms of $\mathbf{b}$ and $\mathbf{c}$ the vectors:
      (i) $\overrightarrow{AP}$ **(ii)** $\overrightarrow{AQ}$ **(iii)** $\overrightarrow{AR}$ **(iv)** $\overrightarrow{PQ}$ **(v)** $\overrightarrow{QR}$
   (b) Show that PQR is a straight line and state the ratio of the lengths $PQ:QR$.

Exercise 25.3   Links: (25G) 25G

1   **(i)** If $3\mathbf{a} = (p - 4)\mathbf{a}$, what is the value of p?
    (ii) If $(2q + 3)\mathbf{b} = 9\mathbf{b}$, what is the value of q?
   (iii) If $(3r - 5)\mathbf{c} = (r + 4)\mathbf{c}$, what is the value of r?

2 Find the magnitude of these vectors
   (a) $\begin{pmatrix} 3 \\ 4 \end{pmatrix}$ **(b)** $\begin{pmatrix} -7 \\ 24 \end{pmatrix}$ **(c)** $\begin{pmatrix} -3 \\ -4 \end{pmatrix}$ **(d)** $\begin{pmatrix} 2 \\ 2 \end{pmatrix}$ **(e)** $\begin{pmatrix} -9 \\ 35 \end{pmatrix}$

3 P is the point $(1, 2)$, Q the point $(3, 1)$ and R is the point $(-2, 0)$.
   (a) Write down the coordinates of the midpoints of
      (i) PQ **(ii)** QR **(iii)** PR
   (b) S lies on PQ extended so that $PS = 2 \times PQ$. Work out the coordinates of S.
   (c) T lies on PR so that $PT = \frac{1}{4}PR$. Work out the coordinates of T.

4 $PQRSTU$ is a regular octahedron. PQ is vector $\mathbf{a}$, QR is vector $\mathbf{b}$, and RT is vector $\mathbf{c}$.
   (a) Write down the vectors
      (i) $\overrightarrow{PT}$ **(ii)** $\overrightarrow{TQ}$ **(iii)** $\overrightarrow{UP}$ **(iv)** $\overrightarrow{SU}$.
   (b) Write down two lines that have the vectors
      (i) $\mathbf{b} + \mathbf{c}$ **(ii)** $\mathbf{a} + \mathbf{c}$ **(iii)** $\mathbf{a} + \mathbf{b} + \mathbf{c}$

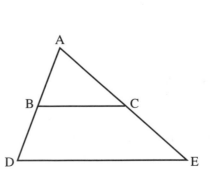

5 In triangle ADE, B and C are the midpoints of AD and AE respectively. If $AB = \mathbf{a}$ and $AC = \mathbf{c}$, prove that BC is parallel to, and half the length of DE.

26 Circle theorems

1 In the diagram, PQ is a chord of the circle centre O.
   The radius of the circle is 17 cm.
   M is the midpoint of PQ.
   $PQ = 30$ cm.
   Calculate the length of OM.

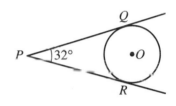

2 PQ and PR are tangents to the circle centre O.
   The angle $QPR = 32°$.
   Calculate the angle POR.

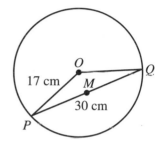

3 AB is a diameter of the circle.
   C is a point on the circumference of the circle.
   $AC = 5$ cm and $BC = 12$ cm.
   Calculate:
   (a) the radius of the circle
   (b) the area of the circle.

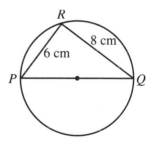

4 PQ is a diameter of the circle.
   R is a point on the circumference of the circle.
   $PR = 6$ cm and $QR = 8$ cm.
   Calculate:
   (a) the circumference of the circle
   (b) the area of the circle.

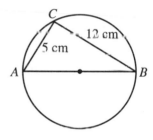

5 AC is a diameter of the circle centre O.
   B is a point on the circumference of the circle.
   P is another point on the circumference of the circle.
   PO is parallel to AB.
   The angle $ACB = x°$
   Find, with reasons and in terms of x, an expression for the angle
   AOP.

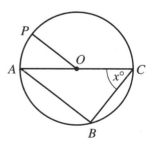

6 *A, B, C* and *D* are four points on the circumference of a circle.
   ABCD is a rectangle. The area of the circle, in terms of π, is
   $400\pi\,\text{cm}^2$.
   Given that the lengths of *AD* and *DC* are integers show that the
   area of *ABCD* is $768\,\text{cm}^2$.

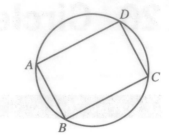

7 *AB* is a diameter of the circle.
   The line *STA* is a tangent to the circle.
   C is a point on the circumference of the
   circle and is such that $TC = CA$.
   The angle $CTS = x°$.
   Find, in terms of *x*, an expression for the angle:
   (a) *TCA*   **(b)** *ABC*

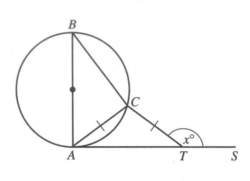

Exercise 26.2   Links: (*28C, D*) 26C, D

1 *P, Q* and *R* are three points on the circle centre *O*.
   Angle $PQR = 54°$.
   Find the angles:
   (a) *POR*   **(b)** *RPO*

   Give your reasons.

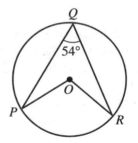

2 The points *A, B* and *C* lie on the circle centre *O*.
   The angle $AOB = 2x°$.
   $AC = BC$.
   Find, in terms of *x*, expressions for the angles:
   (a) *ACB*   **(b)** *ABC*   **(c)** *CBO*

   Give your reasons.

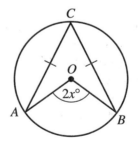

3 *P, Q, R* and *S* are four points on the circle centre *O*.
   The reflex angle $POR = 230°$.
   Calculate the size of the angle:
   (a) *PSR*   **(b)** *PQR*

   Give your reasons.

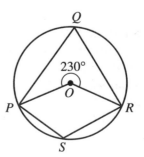

4 The four points A, B, C and D lie on the circle centre O.
Angle $BCD = 48°$.
$AB = AD$.
Calculate the size of the angle:
(a) ADB (b) ODB

Give your reasons.

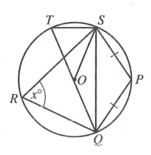

5 P, Q, R and S are four points on the circumference of the circle centre O.
The line segment QOT is a diameter of the circle.
The angle $QRS = x°$.
$PS = PQ$.
Find, in terms of x, expressions for:
(a) the angle QTS (b) the angle QOS (c) the angle QPS
(d) the angle SQP (e) the angle TSO (f) the angle TOS

In the case when the triangle TOS is **equilateral**:
(g) find the value of x.

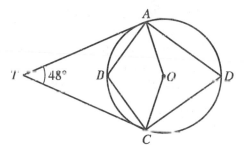

6 A, B, C and D are four points on a circle centre O.
T is a point outside the circle such that TA and TC are tangents to the circle and angle $ATC = 48°$.
Calculate the values of the angles:
(a) AOC (b) ADC (c) ABC

The line AO produced, meets the circle at P.
(d) Calculate the size of the angle APC.

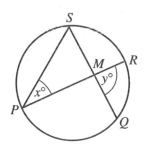

7 P, Q, R and S are four points on the circle centre O.
PR and QS intersect at the point M.
Angle $SPR = x°$ and angle $RMQ = y°$.
Find, in terms of x and/or y, expressions for the angles:
(a) RQS (b) PSQ

Exercise 26.3   Links: (*28E, F*) 26E, F

1 P, Q, R and S are four points on a circle.
T is a point outside the circle with PST a straight line.
Angle $PQR = 48°$
Calculate the size of the angle RST.

Give your reasons.

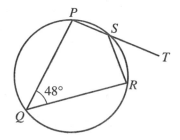

2 *ABCD* is a cyclic quadrilateral.
ABD is an isosceles triangle with *AB* = *AD*.
Angle *ABD* = 52°.
Calculate the size of the angle:
(a) *BAD*   (b) *DCB*

Give your reasons.

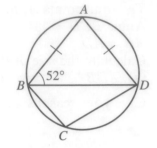

3 *PQRS* is a cyclic quadrilateral.
O is the centre of the circle.
The point *T* lies on the circle and *QOT* is a straight line.
The angle *QTS* = *x*°.
Write down, in terms of *x*, expressions for:
(a) the angle *QPS*   (b) the angle *QOS*   (c) the angle *QRS*

Give your reasons.

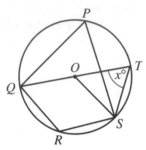

4 The points *A*, *B* and *C* lie on a circle.
TA is a tangent to this circle.
The angle *TAC* = *x*°.
The triangle *ABC* is isosceles with *AB* = *AC*.
Find, in terms of *x*, expressions for the angles:
(a) *ACB*   (b) *BAC*

Give your reasons.

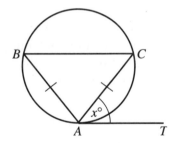

5 *ABCD* is a cyclic quadrilateral.
The line from *D* to the centre of the circle, *O*, is parallel to *AB*.
The angle *DCB* = 62°.
Calculate the angles:
(a) *DOB*   (b) *DAB*   (c) *OBA*

Give your reasons.

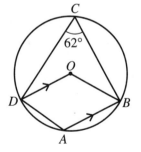

6 The diagram shows two circles which intersect
at the points *Q* and *R*.
PQRS and *NMRQ* are two cyclic quadrilaterals.
P, *Q* and *N* lie on a straight line.
Angle *NRM* = 25°, angle *QNR* = 55° and
angle *QMN* = 53°.
Calculate the sizes of the angles:
(a) *RNM*   (b) *MRQ*   (c) *PSR*

Give your reasons.

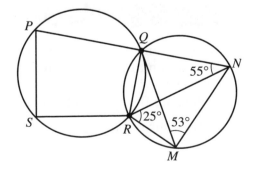

7 *PA* and *PB* are the tangents to a circle, centre *O*, from a point *P* which lies outside the circle. The angle $APB = 50°$. $AB = BC$.

(a) Calculate the sizes of the angles:

　(i) *ACB*　　(ii) *AOC*

(b) Explain why *ACBP* cannot be a cyclic quadrilateral.

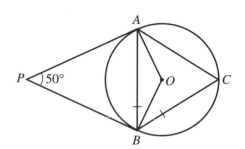

Exercise 26.4　　　　　　　Links: (28G) 26G

1 *AB* and *CD* are two chords of a circle which meet inside the circle at the point *M* which is inside the circle.

(a) Prove that the triangles *MAC* and *MDB* are similar.

(b) Show that $MA \times MB = MD \times MC$.

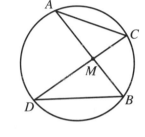

2 *PQ* and *RS* are two chords of a circle which meet at a point *N* which lies outside the circle.

(a) Explain why the triangles *NPS* and *NRQ* are similar.

(b) Prove that $NQ \times NP = NS \times NR$.

$NQ = x$ cm, $QP = 9$ cm, $NS = 4$ cm and $SR = 8$ cm.

(c) Show that $x^2 + 9x - 48 = 0$

(d) Solve the equation to find the value of *x*.

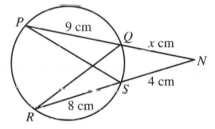

3 The chord *AB* of the circle, when produced, meets the tangent at *C* in the point *T*.

(a) Explain fully, why the triangles *TCA* and *TBC* are similar.

(b) Show that:

　(i) $\dfrac{TC}{TB} = \dfrac{TA}{TC}$　　(ii) $TC^2 = TA \times TB$

$TA = x$ cm, $AB = 6$ cm and $TC = 4$ cm.

(c) Prove that $x^2 - 6x - 16 = 0$

(d) Solve this equation to find the value of *x*.

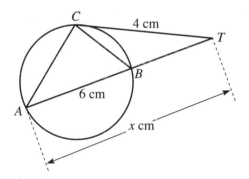

4 *PA* and *PB* are tangents to a circle centre *O*, which meet at *P*. *PO* produced meets the circle at *C*. Angle $APO = x°$.

Explain fully why angle $ACP =$ angle $BCP = \frac{1}{2}(90 - x)°$.

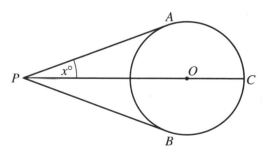

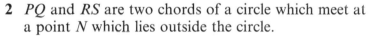

5 The tangent at T to the circle is parallel to the chord AB.
Prove that the triangle TAB is isosceles.

6 Two circles, centres O and Q touch at T.
The common tangent at T meets the other common
tangents AB and CD at X and Y respectively.
(a) Explain why $AOTX$ is a cyclic quadrilateral.
(b) Explain why angle $BXT =$ angle AOT.

7 Two circles touch externally
at a point M.
A line through M meets the circles
at points A and B.
Explain why the tangent at A and
the tangent at B **must** be parallel.

Exercise 26.5   **Links:** (*28H, I*) **26H, I**

1 The diagram shows a circle centre O. PQ and RQ are
tangents to the circle at P and R respectively.
S is a point on the circle.
Angle $PSR = 70°$.
$PS = SR$.
  (a) **(i)** Calculate the size of the angle PQR.
        (ii) State the reasons for your answer.
  (b) **(i)** Calculate the size of the angle SPQ.
        (ii) Explain why $PQRS$ cannot be a cyclic quadrilateral. [L]

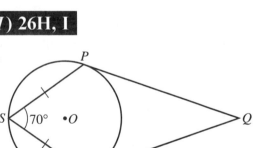

2 A, B and C are three points on the circle.
Each of the angles in the triangle ABC is acute.
The tangents at A and C meet at T.
M is the point on AT such that angle $ACM =$ angle TCM.
Angle $CMT = 81°$.
Giving all of your reasons, calculate the size of the angle ABC.

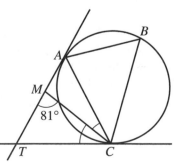

3 The points A, B and C lie on a circle.
PAQ is a tangent to this circle.
PQ is parallel to BC. Angle $PAB = x°$.
Find, showing your reasons, an expression in terms of x for the angle BAC.

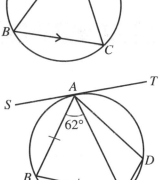

4 $ABCD$ is a cyclic quadrilateral.
SAT is a tangent at A.
$AB = BC$. $BAC = 62°$.
Find, showing all your reasons:
  (a) angle BCA **(b)** angle SAB **(c)** angle ADC

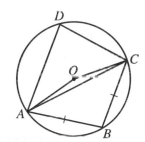

5 $ABCD$ is a cyclic quadrilateral. O is the centre of the circle.
$AB = BC$.
  (a) Prove that angle $ADC = 2 \times$ angle ACB.

Angle $BAC = x°$.
  (b) Write down an expression for the angle AOC.

6 Chords AB and CD meet at the point P outside the circle.
  (a) Write down, with reasons, a triangle similar to APD.

$PB = x$ cm, $BA = 4$ cm, $PD = 5$ cm and $DC = 7$ cm
  (b) Prove that $x^2 + 4x - 60 = 0$
  (c) Find the value of x.

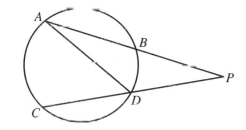

7 AT and BT are two tangents of a circle centre O, which meet at the point T.
$AC = BC$ and angle $TAC = x°$.
  (a) Find, in terms of x, expressions for:
    (i) angle ACB **(ii)** BOA
  (b) Explain whether or not the quadrilateral $AOBC$ can be cyclic.

8 A, B, C, are points on a circle, centre O.
TA is a tangent.
Angle $C\widehat{A}T = x°$.
Without stating the alternate segment theorem, prove that angle $A\widehat{B}C = x°$.

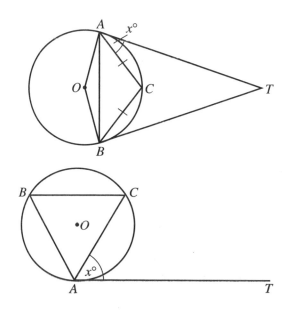

27 Histograms and dispersion

1 The heights in metres of 30 sunflowers are shown below:

2.76	3.45	2.91	1.78	2.90	2.46	3.04	2.87	2.68	3.15
2.26	1.98	3.79	2.86	2.49	2.46	2.76	3.12	3.89	4.20
1.98	3.12	2.56	2.68	2.42	2.34	2.78	3.01	1.46	2.73

(a) Draw up a frequency table with class intervals of 1.0 to 2.0, 2.0 to 2.5, 2.5 to 2.75, 2.75 to 3.25, 3.25 to 4.0, 4.0 to 4.5.

(b) Draw a histogram to show this data.

2 The time to cycle 1000 m was measured for members of a cycle club. The times taken were recorded in a table.

Time to cycle 1000 m (s)	Frequency
120–150	6
150–170	15
170–210	18
210–250	9
250–300	2

Draw a histogram to show this data.

3 This histogram represents the lengths of cuttings planted by a gardener one weekend.

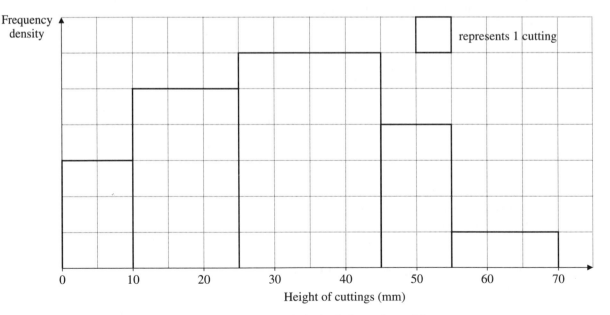

(a) Write down the number of cuttings with length less than 25 mm.
(b) Write down the number of cuttings with length greater than or equal to 10 mm and less than 55 mm.
(c) Calculate the total number of cuttings in the sample.

4 This histogram represents the number of spectators at professional rugby matches on one Saturday in 1998.

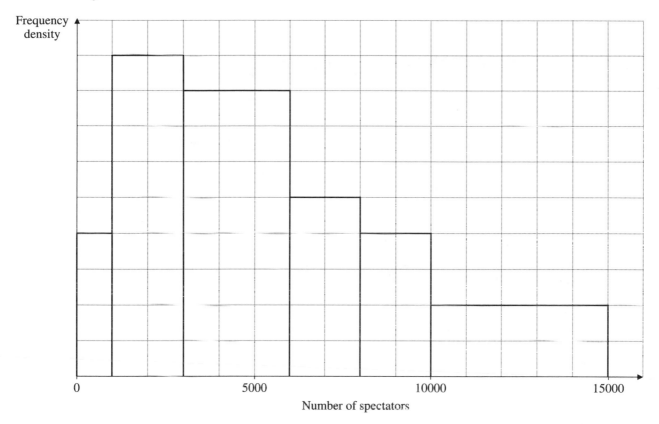

No match had more than 15 000 spectators. At 5 matches the number of spectators was greater than or equal to 6000 and less than 8000.

(a) Use the above information to complete this frequency table:

(b) Calculate the total numbers of professional rugby matches played on that Saturday.

Number of spectators (n)	Frequency
$0 \leqslant n < 1000$	
$1000 \leqslant n < 3000$	
$3000 \leqslant n < 6000$	
$6000 \leqslant n < 8000$	5
$8000 \leqslant n < 10\,000$	
$10\,000 \leqslant n < 15\,000$	

5 This table shows the distribution of weight, in kilograms of 40 sheep at Lakes Farm.

Draw a histogram to represent this data.

Weight w (kg)	Frequency
$20 \leqslant w < 25$	2
$25 \leqslant w < 35$	9
$35 \leqslant w < 40$	10
$40 \leqslant w < 50$	12
$50 \leqslant w < 65$	2
$65 \leqslant w < 80$	1

28 Introducing modelling

1 George bought a new car on 1st August 1997
for £12 000.
The value of the car **depreciates** by
15% each year.

(a) Copy and complete this table for the value
of the car, giving the values correct to the
nearest £10.

Year on August 1st	1997	1998	1999	2000	2001	2002
Value (£)	12 000					

(b) Draw a graph of the value of the car against the year from
1997 to 2002.
(c) Find a formula which models the value of the car *n* years
after it was bought new.
(d) Use your formula to work out:
   (i) the value of the car on 1st August 2007
   (ii) the year in which the value of the car first falls
        below £500.

2 Julie started to work for a jeweller on 1st January 1998.
Her wage was fixed at £200 per week.
The jeweller promised Julie a pay rise of 5% per year.
If Julie stays with the jeweller, what could be her expected
weekly pay after she has been working there for:
(a) 2 years   (b) 8 years   (c) *n* years?

3 *t* hours after midnight one day, the depth of water,
d metres, at the entrance to a small fishing harbour
is modelled by the formula:

$$d = 8 + 5\sin(20t)°$$

(a) Calculate the depth of water at:
   (i) 2 am   (ii) 6 am   (iii) 10 am
(b) Work out the lowest value for the depth of the water
at the entrance to the harbour.
(c) Work out the largest value of the depth of the water
at the entrance to the harbour.
(d) Find the times of low tide and high tide during the day.
(e) Sketch a graph of the depth of the water at the entrance to
the harbour as the time varies from 0 to 24 hours.

4 The diagram represents a moving particle *P* and a fixed point *O*.

$$\begin{array}{ccc} & O & P \\ \rule{3cm}{0.4pt}\!\!\bullet\!\!\rule{2cm}{0.4pt}\!\!\bullet\!\!\rule{1.5cm}{0.4pt} \end{array}$$

The particle moves in a straight line such that its distance from *O*, *y* metres, at any time *t* seconds is given by the formula:

$$y = a + b\cos(30t)^\circ$$

When $t = 0$, $y = 3$
When $t = 2$, $y = 4$.

(a) Show that:

$$a + b = 3$$
$$2a + b = 8$$

(b) Solve these equations to find the values of *a* and *b*.
(c) Sketch the graph of *y* against *t* for values of *t* from 0 to 10.
(d) Find the value of *y* when:
    (i) $t = 2$ **(ii)** $t = 5$
(e) Find the maximum distance between *O* and *P* and the values of *t* when this occurs.

Exercise 28.2   Links: (*30C, D*) 28C, D

1 The table below contains information about the diameters, in centimetres and heights, in metres, of ten horse chestnut trees.

Diameter (cm)	120	140	141	132	134	130	112	150	147	122
Height (m)	27	33	31	30	31	32	28	33	31	30

(a) Plot these points on a scatter diagram.
(b) Draw a line of best fit on your scatter diagram.
(c) Find the equation of your line of best fit.
(d) Use the equation of the line to work out estimates of:
    (i) the height of a horse chestnut tree of diameter 135 cm
    (ii) the diameter of a horse chestnut tree of height 29 m.

2 Repeat parts **(a)**, **(b)** and **(c)** above for the heights and diameters of these beech trees.

Diameter (cm)	170	160	183	162	164	180	170	172	194	203	210	201
Height (m)	36	32	39	32	33	37	32	34	37	39	41	40

3 In an experiment, two variables, x and y are thought to be
   connected by a relationship of the type:

   $$y = ax^2 + b$$

   where a and b are constants.
   Corresponding values of x and y are given in the table below.

x	1	1.5	2	2.3	3.5	4
y	-1	1.5	5	7.58	21.5	29

   (a) By plotting a suitable straight line graph, show that x and y
        are related by the given type of relationship.
   (b) Use your graph to work out the values of a and b.
   (c) Work out the values of:
        (i) x when $y = 3$ **(ii)** y when $x = 2.4$ **(iii)** x when $y = 19.5$

4 The table shows the stopping distance, d feet,
   for a car travelling at a speed of s mph.

s	20	30	40	50	60	70
d	40	75	120	175	240	315

   (a) Copy and complete this table of values for $\dfrac{d}{s}$ against s.

s	20	30	40	50	60	70
$\dfrac{d}{s}$	2	2.5				

   (b) Plot the graph of $\dfrac{d}{s}$ against s for values of s from 20 to 70.

   (c) Explain fully why this graph confirms that d and s are
        connected by a relationship of the type:

        $$d = as^2 + bs$$

        in which a and b are constants.
   (d) Use your graph to work out the values of a and b.
   (e) Without using your graph, work out the value of:
        (i) the stopping distance for a car travelling at a speed of
             35 mph.
       (ii) the speed a car was travelling at when it is known that
             its stopping distance was 300 feet
      (iii) the stopping distance for a car travelling at 100 mph.

   At the scene of an accident, skid marks indicate that the
   stopping distance for a car was 145 feet.
   The maximum speed limit on the road is 45 mph.
   (f) Explain whether or not the skid marks provide evidence that
        the car was breaking the speed limit just prior to the
        accident.

5 Shivana has been conducting an experiment in science.
She believes that two variables y and t are connected
by a formula of the type:

$$y = at^2 + bt$$

She has five sets of results from her experiment.
These results appear in the table below.

t	1	2	3	4	5
y	5.1	8.3	9.3	8.1	5.1

(a) By drawing a suitable straight line graph, show that,
within experimental error, the results confirm Shivana's
believed formula.

(b) Use your graph to find the values of a and b.

Shivana will test her formula by repeating the experiment for a
sixth time using $t = 4.5$.

(c) Use the formula to work out an approximate value for y in
this case.

Exercise 28.3   Links: (*30E*) 28E

1 This sketch shows part of the graph of:

$$y = pq^x$$

It is known that the points $(0, 3)$, $(2, k)$ and $(4, 1875)$ lie on the
curve.
Use the sketch and the given information to find the values of
p, q and k.

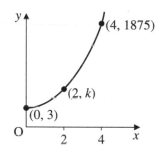

2 The point $(1, 16)$ lies on the curve $y = a^{2x}$.
Calculate the values of a.

3 The diagram represents a sketch of part of the graph of:

$$y = p + q^x$$

Use the sketch to find the values of p, q and k.

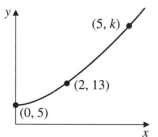

4 The time, t seconds, for one complete oscillation of a pendulum
of length l metres is given by the formula:

$$t = kl^n$$

When $l = 1$, $t = 2$ and when $l = 9$, $t = 6$.
Work out the value of n.

Heinemann Educational Publishers
Halley Court, Jordan Hill, Oxford, OX2 8EJ
a division of Reed Educational & Professional Publishing Ltd
Heinemann is a registered trademark of Reed Educational & Professional
Publishing Ltd

OXFORD MELBOURNE AUCKLAND
JOHANNESBURG BLANTYRE GABARONE
IBADAN PORTSMOUTH (NH) USA CHICAGO

Gareth Cole, David Kent, Peter Jolly, Keith Pledger, 1998, 2002

First published 2002

ISBN 0 435 53268 5

06 05 04 03 02
10 9 8 7 6 5 4 3

Designed and typeset by Techset, Tyne and Wear

Cover design by Miller, Craig and Cocking

Printed and bound by The Bath Press, Bath

Acknowledgements

The publisher's and author's thanks are due to Edexcel Foundation for permission
to reproduce questions from past examination papers. These are marked with an
[E]. The answers have been provided by the authors and are not the responsibility
of Edexcel.